Control System Synthesis

A Factorization Approach

Part I

Synthesis Lectures on Control and Mechatronics

Editors
Chaouki Abdallah, *University of New Mexico*
Mark W. Spong, *University of Texas at Dallas*

Control System Synthesis: A Factorization Approach, Part I
Mathukumalli Vidyasagar
2011

The Reaction Wheel Pendulum
Daniel J. Block, Karl J. Åström, and Mark W. Spong
2007

Control System Synthesis: A Factorization Approach, Part I
Mathukumalli Vidyasagar

ISBN: 978-3-031-00700-2   paperback
ISBN: 978-3-031-01828-2   ebook (Part I)

DOI 10.1007/978-3-031-01828-2

A Publication in the Springer series
SYNTHESIS LECTURES ON CONTROL AND MECHATRONICS

Lecture #2
Series Editors: Chaouki Abdallah, *University of New Mexico*
                Mark W. Spong, *University of Texas at Dallas*
Series ISSN
Synthesis Lectures on Control and Mechatronics
Print 1939-0564   Electronic 1939-0572

Control System Synthesis

A Factorization Approach

Part I

Mathukumalli Vidyasagar
University of Texas at Dallas

SYNTHESIS LECTURES ON CONTROL AND MECHATRONICS #2

ABSTRACT

This book introduces the so-called "stable factorization approach" to the synthesis of feedback controllers for linear control systems. The key to this approach is to view the multi-input, multi-output (MIMO) plant for which one wishes to design a controller as a matrix over the fraction field **F** associated with a commutative ring with identity, denoted by **R**, which also has no divisors of zero. In this setting, the set of single-input, single-output (SISO) stable control systems is precisely the ring **R**, while the set of stable MIMO control systems is the set of matrices whose elements all belong to **R**. The set of unstable, meaning not necessarily stable, control systems is then taken to be the field of fractions **F** associated with **R** in the SISO case, and the set of matrices with elements in **F** in the MIMO case.

The central notion introduced in the book is that, in most situations of practical interest, every matrix P whose elements belong to **F** can be "factored" as a "ratio" of two matrices N, D whose elements belong to **R**, in such a way that N, D are coprime. In the familiar case where the ring **R** corresponds to the set of bounded-input, bounded-output (BIBO)-stable rational transfer functions, coprimeness is equivalent to two functions not having any common zeros in the closed right half-plane including infinity. However, the notion of coprimeness extends readily to discrete-time systems, distributed-parameter systems in both the continuous- as well as discrete-time domains, and to multi-dimensional systems. Thus the stable factorization approach enables one to capture all these situations within a common framework.

The key result in the stable factorization approach is the parametrization of *all* controllers that stabilize a given plant. It is shown that the set of all stabilizing controllers can be parametrized by a single parameter R, whose elements all belong to **R**. Moreover, every transfer matrix in the closed-loop system is an affine function of the design parameter R. Thus problems of reliable stabilization, disturbance rejection, robust stabilization etc. can all be formulated in terms of choosing an appropriate R.

This is a reprint of the book *Control System Synthesis: A Factorization Approach* originally published by M.I.T. Press in 1985.

KEYWORDS

stable factorization, coprimeness, Bezout identity, simultaneous stabilization, robust stabilization, robust regulation, genericity

Dedicated to the memories of
My paternal grandfather,
Mathukumalli Narasimharao (1881–1965)
and
My father,
Prof. Mathukumalli Venkata Subbarao (1921–2006)

ॐ सह नाववतु
सह नौ भुनक्तु
सह वीर्यम् करवावहै
तेजस्वि नावधीतमस्तु
मा विद्विषावहै
ॐ शंतिः शंतिः शंतिः

May God protect us together
May God nourish us together
May we work together with great energy
May our study be vigorous and effective
May we not have ill feelings
Let there be peace, peace, peace

(A vedic prayer recited by the teacher and student(s) together at the commencement of studies. Found in several upanishads including *Taittiriyopanishad*)

Contents

Preface

This is a reprint of my book *Control System Synthesis: A Factorization Approach*, originally published by M.I.T. Press in 1985. That book went out of print about ten years after its publication, but nevertheless continued to be cited in the published literature. I would like to believe that this is because the contents of the book are still relevant to linear control theory as it is currently practiced. I am therefore grateful to Morgan & Claypool for their decision to reprint my book, so that the current generation of graduate students and researchers are able to access its contents without having to photocopy it surreptitiously from the library.

The original text was created by me using the troff text processing system created at Bell Laboratories. Indeed, the 1985 book was the first time that I produced a camera-ready copy, instead of subjecting myself to the mercy and vagaries of typists and/or typesetters. Back then this was a novel thing to do; however today it is standard practice. Dr. C. L. Tondo of T&T TechWorks, Inc. and his troops have expertly keyed in the entire text into LATEX, for which I am extremely grateful. There is no doubt that the physical appearance of the text has been significantly improved as a result of this switch.

With the entire text at my disposal, I could in principle have made major changes. After thinking over the matter, I decided to stick with the original text, and restricted myself to correcting typographical errors. Upon re-reading the original text after a gap of perhaps twenty years, I felt that the entire material still continues to be relevant, except for Chapter 6 on H_∞-control. Just about the time that I was finalizing the book, two seminal papers appeared, giving the connection between interpolation theory and H_∞-control; these are cited in the book as references [40] and [115]. A few years after my book was published, the influential book [37] appeared. This was followed in short order by the paper [28], which gives a complete state-space computational procedure for H_∞-control; this paper is perhaps the most cited paper in the history of control theory. Subsequently, the book [29] gave an elementary introduction to the theory, while [45, 117] are advanced treatments of the subject. It would therefore have required a massive effort on my part to rewrite Chapter 6 of the book to bring it up to date, and I felt that I could contribute nothing beyond the excellent texts already in print. So I decided to leave the book as it is, on the basis that the conceptual framework for H_∞-control presented here still remains relevant.

I had dedicated the original book to my paternal grandfather. In the interim, my father too has passed away, and I have therefore added his name to the dedication of the Morgan & Claypool edition. About twenty years ago, while perusing a book on Telugu writers, I discovered that one of my ancestors, bearing the same name as my paternal grandfather, was a well-known Telugu poet during the 19th century; he lived between 1816 and 1873. The article about him states that the clan was famous for scholarship for *three centuries* (emphasis added). I am truly blessed to have such a distinguished lineage.

Dallas and Hyderabad, June 2011

Preface for the Original Edition

The objective of this book is to present a comprehensive treatment of some recent results in the area of linear multivariable control that can be obtained using the so-called "factorization" approach. It is intended as a second level graduate text in linear systems and as a research monograph. The prerequisites for reading this book are covered in three appendices, but a reader encountering these topics for the first time would undoubtedly have difficulty in mastering this background on the basis of these appendices alone. Moreover, the appendices concentrate on the mathematical background needed to understand the material covered here, but for motivational background a standard first course in graduate level linear system theory would be desirable.

The central idea that is used repeatedly in the book is that of "factoring" the transfer matrix of a (not necessarily stable) system as the "ratio" of two stable rational matrices. This idea was first used in a paper published in 1972 (see [92]), but the emphasis there was on analyzing the stability of a given system rather than on the synthesis of control systems as is the case here. It turns out that this seemingly simple stratagem leads to conceptually simple and computationally tractable solutions to many important and interesting problems; a detailed description can be found in Chapter 1.

The starting point of the factorization approach is to obtain a simple parametrization of all compensators that stabilize a given plant. One could then, in principle, choose the best compensator for various applications. This idea was presented in the 1976 paper by Youla, Jabr, and Bongiorno entitled "Modern Wiener-Hopf Design of Optimal Controllers, Part II: The Multivariable Case," which can be considered to have launched this entire area of research. The viewpoint adopted in this book, namely that of setting up all problems in a ring, was initially proposed in a 1980 paper by Desoer, Liu, Murray, and Saeks. This paper greatly streamlined the Youla et al. paper and reduced the problem to its essentials. Thus virtually all of the research reported here is less than five years old, which bears out the power of this approach to formulate and solve important problems.

In writing the book, some assumptions have been made about the potential readership. First, it is assumed that the reader is already well versed in the aims and problems of control system analysis and design. Thus, for example, the book starts off discussing the problem of stabilization without any attempt to justify the importance of this problem; it is assumed that the reader already knows that stable systems are better than unstable ones. Also, as the book is aimed at professional researchers as well as practitioners of control system synthesis, a theorem-proof format has been adopted to bring out clearly the requisite conditions under which a particular statement is true, but at the same time, the principal results of each section have been stated as close to the start of the section as is practicable. In this way, a reader who is interested in pursuing the topics presented here in greater depth is enabled to do so; one who is only interested in using the results can rapidly obtain an idea of what they are by scanning the beginnings of various sections and by skipping proofs. In this connection it is worth noting that Chapter 3 is a painless introduction to the factorization approach to scalar systems that could, in my opinion, be taught to undergraduates without difficulty.

At various times, I have taught the material covered here at Waterloo, Berkeley, M.I.T., and the Indian Institute of Science. Based on these experiences, I believe that the appendices plus the first five chapters can be covered in a thirteen-week period, with three hours of lectures per week. In a standard American semester consisting of eighteen weeks of lectures with three hours of lectures per week, it should be possible to cover the entire book, especially if one starts directly from Chapter I rather than the appendices. Most of the sections in the appendices and the first five chapters contain several problems, which contain various ancillary results. The reader is encouraged to attempt all of these problems, especially as the results contained in the problems are freely used in the subsequent sections.

It is now my pleasure to thank several persons who aided me in this project. My wife Shakunthala was always a great source of support and encouragement during the writing of this book, which took considerably more time than either of us thought it would. Little Aparna came into being at about the same time as the third draft, which gave me a major incentive to finish the book as quickly as possible, so that I might then be able to spend more time with her. Several colleagues gave me the benefit of their helpful comments on various parts of the manuscript. Of these, I would like to mention Ken Davidson, Charlie Desoer, John Doyle, Bruce Francis, Allen Tannenbaum, George Verghese, N. Viswanadham, and Alan Willsky. My students Chris Ma and Dean Minto went over large parts of the material with a great deal of care and exposed more than one serious mistake. I would like to acknowledge my indebtedness to all of these individuals. The final camera ready copy was produced by me using the troff facility and the pic preprocessor to generate the diagrams. In this connection, I would like to thank. Brian Haggman, a differential topologist turned computer hack, for much valuable advice. Finally, at the outset of this project, I was fortunate enough to receive an E. W. R. Steacie Memorial Fellowship awarded by the Natural Sciences and Engineering Research Council of Canada, which freed me from all teaching duties for a period of two years and enabled me to concentrate on the research and writing that is reflected in these pages.

I would like to conclude this preface with a historical aside, which I hope the reader will find diverting. It is easy enough to discover, even by a cursory glance at the contents of this book, that one of the main tools used repeatedly is the formulation of the general solution of the matrix equation $XN + YD = I$, where all entities are matrices with elements in a principal ideal domain. Among recent writers, Tom Kailath [49] refers to this equation as the Bezout identity, while V. Kučera [60] refers to it as the Diophantine equation. In an effort to pin down just exactly what it should be called, I started searching the literature in the area of the history of mathematics for the details of the person(s) who first obtained the general solution to the equation $ax + by = 1$, where a and b are given integers and one seeks integer-valued solutions for x and y.

It appears that the equation "Diophantine equation" was commonly applied by European mathematicians of the seventeenth century and later to any equation where the unknowns were required to assume only integer values. The phrase honors the Greek mathematician Diophantus, who lived (in Alexandria) during the latter part of the third century A. D. However, the general solution of the linear equation in integer variables mentioned above was never studied by him. In fact, Smith [90, p. 134] states that Diophantus never studied indeterminate equations, that is equations that have more than one solution. According to Colebrooke [17, p. xiii], the first occidental mathematician to study this equation and to derive its general solution was one Bachet de Meziriac in the year 1624. The first mathematician of antiquity to formulate and find all solutions to this equation was an ancient Hindu by the name of Aryabhatta, born in the year 476. A detailed and enjoyable exposition of this subject can be found in the recent comprehensive book by van der Waerden [104]. Thus, in order to respect priority, I submit that the equation in question should henceforth be referred to as Aryabhatta's identity.

CHAPTER 1

Introduction

The objective of this book is to present a comprehensive treatment of various recent results in the area of linear multivariable control that can be obtained using the so-called *factorization approach*. The central idea of this approach is to "factor" the transfer matrix of a (not necessarily stable) system as the "ratio" of stable matrices. This seemingly simple idea gives rise to an elegant methodology which leads in a simple and natural way to the resolution of several important control problems. Moreover, the approach given here is very general, encompassing continuous-time as well as discrete-time systems, lumped as well as distributed systems, and one-dimensional as well as multidimensional systems, all within a single framework.

The basic problem studied in this book can be stated as follows: Given a plant P, which is assumed to be linear and time-invariant (or, in the case of multi-dimensional systems, shift-invariant), together with a set of performance specifications, design a compensator C such that the plant-compensator pair meets the performance requirements. The performance requirements studied can be classed into two types: (i) The transfer matrix of the compensated system is "desirable" in some sense to be made precise shortly. (ii) The transfer matrix of the compensated system is optimal in some sense. To put it another way, the class of compensator design problems studied here can be classed into two categories: In the first category, one is given a plant, together with a specification of a class of desirable transfer functions (or matrices), and any compensator is deemed to be acceptable so long as it results in a compensated system whose transfer matrix lies within the desirable class. A typical example of this sort of problem is that of stabilizing an unstable system. In the second category, one is given a plant, a class of desirable transfer matrices, as well as a performance measure, and the objective is to choose, among all the compensators that result in a desirable closed-loop transfer matrix, one that optimizes the performance measure. Illustrations of this class of problems include those of designing a compensator for a system so as to achieve optimal disturbance rejection, optimal robustness, optimal tracking etc., all while maintaining closed-loop stability.

Within the first class of problems, the important specification is that of the class of transfer functions that are deemed to be "desirable." This class can vary from one application to another. For instance, in the case of lumped continuous-time systems, if one is interested in the bounded-input-bounded-output (BIBO) stability of the compensated system, then the set of desirable transfer functions consists of all proper rational functions in the Laplacian variable s whose poles all lie in the open left half-plane. However, in many applications, mere BIBO stability is not enough, and it is required that the poles of the transfer matrix of the compensated system lie in some more restrictive region of stability $\mathbf{D}$, as shown in Figure 1.1.

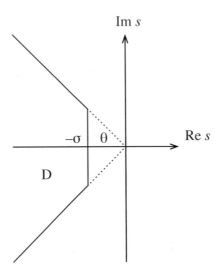

Figure 1.1: Generalized Region of Stability.

If this requirement is satisfied, then the step response of the compensated system exhibits a settling time of no more than $4/\sigma$ and a maximum overshoot corresponding to the angle θ. In the case of discrete-time systems, the stable region is the exterior of the closed unit disc, and the set of "stable" transfer functions are those rational functions that are analytic over the closed unit disc. If a more stringent form of stability is required, the "forbidden region" where the poles of desirable transfer functions may not lie can be enlarged.

In all of the above illustrations, there is one constant feature, namely: cascade and parallel connections of desirable systems are themselves desirable. In more mathematical terms, this can be restated as follows: Let **R** denote the set of desirable transfer functions in a particular application. Then sums and products of functions in the set **R** once again belong to **R**. To put it another way, the set **R** is a *ring* (see Appendix A for a precise definition). This now leads to a slightly more formal statement of the first category of compensator design problems, which is referred to hereafter as the *stabilization problem*: Given a plant P, together with a ring **R** of stable transfer functions,[1] find a compensator C such that the plant-compensator pair (P, C) is stable. There are several possible compensation schemes, but it is by now well-known that the feedback type of compensation is best from the viewpoints of sensitivity reduction, robustness against modeling uncertainty, etc. There are two types of feedback systems studied in this book, namely the "one-parameter" scheme shown in Figure 1.2 and the "two-parameter" scheme shown in Figure 1.3. (The rationale behind the names is explained in Chapter 5.)

---

[1]It is much more evocative to use the term "desirable transfer function" rather than "stable" transfer function, but the use of the latter term is by now entrenched.

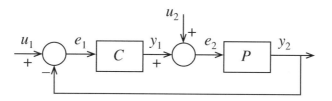

Figure 1.2: One-Parameter Compensation Scheme.

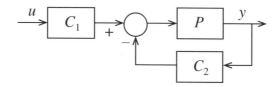

Figure 1.3: Two-Parameter Compensation Scheme.

In the above problem statement, we have specified that the set **R** of desirable transfer functions is a ring, but have not specified the class of transfer functions within which the plant transfer matrix P may lie. To put it in other words, we have specified that the universe of "stable" transfer functions must be a ring, but have not specified the universe of "unstable" (meaning not necessarily stable) systems. Actually, this latter class is specified automatically as a by-product of choosing the compensation scheme.

Consider the feedback configuration shown in Figure 1.2. The transfer matrix from the two external inputs u_1 and u_2 to the outputs y_1 and y_2 of the compensator and plant, respectively, is

$$W(P, C) = \begin{bmatrix} C(I + PC)^{-1} & PC(I + PC)^{-1} \\ CP(I + CP)^{-1} & P(I + CP)^{-1} \end{bmatrix}. \tag{1.0.1}$$

This formula can be written in a more compact form. Define

$$G = \begin{bmatrix} C & 0 \\ 0 & P \end{bmatrix}, \quad F = \begin{bmatrix} 0 & I \\ -I & 0 \end{bmatrix}. \tag{1.0.2}$$

Then

$$W(P, C) = G(I + FG)^{-1}. \tag{1.0.3}$$

Now, as discussed above, the system of Figure 1.2 is "stable" (we drop the quotation marks hereafter) if every element of the matrix $W(P, C)$ belongs to the specified ring **R** of stable transfer functions. Let W denote the matrix $W(P, C)$, and solve (1.0.3) for G. This gives

$$\begin{aligned} G &= W(I - FW)^{-1} \\ &= W(I - FW)^{adj}/|I - FW|, \end{aligned} \tag{1.0.4}$$

where M^{adj} and $|M|$ denote respectively the adjoint and the determinant of a square matrix M. The last expression in (1.0.4) shows that every element of the matrix G can be expressed as a ratio of two functions belonging to **R**. In particular, every element of the matrix P can be expressed as a ratio of two functions in **R**. Thus the conclusion is that, once a set **R** of stable transfer functions has been specified, the class of all (possibly unstable) transfer functions that can be encompassed within the theory of stabilization consists of those that can be expressed as a *ratio* of stable transfer functions.

This last observation leads to a final and precise statement of the stabilization problem, as follows: Given a ring **R** of stable transfer functions, together with a plant transfer matrix P where each element of P is a ratio of functions in **R**, find *all* compensators C that result in a closed-loop transfer matrix whose elements all belong to **R**.

The stabilization problem as stated above differs in a fundamental manner from those treated in earlier design theories such as LQG (linear-quadratic-Gaussian), INA (inverse Nyquist array), CL (characteristic loci), etc. The latter techniques are all addressed to producing *several*, but by no means *all*, compensators that stabilize a given plant. In contrast, the present problem has as its objective the determination of *all* stabilizing compensators corresponding to a particular plant. The reasons for formulating the problem in this manner become clear shortly.

We are now in a position to describe concisely the contents of the book. Chapter 2 is devoted to a study of the set **S**, consisting of all proper rational functions whose poles are all in the open left half-plane. This is precisely the set of transfer functions of BIBO-stable lumped continuous-time systems. By examining various features of this set, such as its algebraic properties and a suitable definition of a norm, we are then able to give in Chapter 3 a quick introduction to the factorization approach for the special case of single-input-single-output systems. It is by now widely recognized that the design of multivariable control systems poses challenges that are fundamentally different in nature from those encountered in the design of scalar systems. The techniques required to extend the simple methods of Chapter 3 to multivariable systems are presented in Chapter 4, which deals with various properties of matrices whose elements all belong to a ring.

With all of this background material in hand, we then proceed to the main body of the book, which commences with Chapter 5. This chapter is addressed to the stabilization problem stated above. The first principal result of this chapter is given in Section 5.2, and consists of a simple parametrization of all compensators that stabilize a given plant, together with a corresponding parametrization of all stable transfer matrices that can be obtained as a result of stabilizing this plant. The utility of this result lies in the fact that both sets can be parametrized in terms of a single "free" parameter which ranges over all matrices (of appropriate dimensions) whose elements are stable functions, and that the stable closed-loop transfer matrix corresponding a particular compensator is an *affine* function of this parameter. This parametrization can be used to solve a variety of design problems where one is required to find a stabilizing compensator that also satisfies some additional constraints. Two typical problems are mentioned here, namely those of reliable stabilization, and regulation. *Reliable stabilization* refers to a situation where it is desired to maintain closed-loop stability in the face of plant/controller failures; this is referred to elsewhere as the problem of designing control systems with "integrity." For example, suppose one is provided with a transfer matrix P_0 that describes the plant under study during normal operating conditions, together with a set $P_1, \ldots, P_l$ of transfer matrices that describe the same plant in various failed modes, e.g., the loss of sensors and/or actuators, changes in operating point, etc. The problem of finding a compensator that stabilizes this plant reliably against the specified list of contingencies is mathematically one of finding a compensator that stabilizes *each* of the plants $P_0, \ldots, P_l$. This is called *simultaneous stabilization*, and is solved in Section 5.4. *Regulation* refers to a situation where a compensated plant is not only required to be stable, but also to track a given reference signal (such as a step, sinusoid, etc.) or to reject a disturbance signal occurring at the output of the plant or compensator. This problem need not always have a solution, and conditions for it to be solvable are given in Section 5.7. In both of the above applications, as well as others that are studied in Chapter 5, the parametrization given in Section 5.2 plays a central role, especially in concluding the nonexistence of a solution. In fact, problems such as simultaneous stabilization are almost impossible to analyze without a knowledge of *all* stabilizing compensators corresponding to a given plant, while the regulation problem is reduced to the analysis of a linear equation over a ring with the aid of the parametrization of all stable transfer matrices obtainable from a given plant.

Chapter 6 is addressed to problems of filtering and sensitivity minimization. The theme here is that one is interested in selecting, among all compensators that stabilize a given plant, one that is optimal in some sense. The criterion of optimality used here is the ability of the compensated system to track a (stable) reference signal, or to reject a (stable) disturbance signal. Depending on the norm used to measure the tracking error, one is led to two distinct classes of mathematical problems, these being filtering and sensitivity minimization. The filtering problem is closely related to the classical Wiener-Hopf problem. The sensitivity minimization problem is very recent in origin, and is also referred to as $\mathbf{H}_\infty$–norm minimization.

Chapter 7 is devoted to a study of the robustness of stabilization. This question arises because there are always errors and/or simplifications in formulating the model of the plant to be stabilized, and in implementing the compensator. Thus it is important to have an idea of the nature and extent of the uncertainties in the plant and/or compensator that can be permitted without destroying the stability of the compensated system. This issue is fully resolved in this chapter. In addition, the problems of optimal tracking and disturbance rejection are revisited, this time without the assumption that the signals being tracked or rejected are stable.

In the last chapter of the book, some of the contents of the preceding chapters are extended to more general situations. Once one leaves the realm of lumped systems (be they continuous-time or discrete-time), the mathematical structure of the underlying set of stable functions becomes much more complicated. As a consequence, some of the results of the earlier chapters do not carry over smoothly to the more general case. Several do, however, and this chapter contains an exposition of these.

In conclusion, this book presents some significant contributions to the *theory* of linear multi-variable control, which make it possible to resolve several important synthesis problems in a logically simple manner. The next step is to develop computationally efficient and numerically robust algorithms to carry out the various procedures advanced here. Much work has already been done in this direction, and once this task is completed, the factorization approach will provide an attractive alternative to existing design methodologies.

CHAPTER 2

Proper Stable Rational Functions

In this chapter, we study the ring of proper stable rational functions, which plays a central role in the synthesis theory developed in this book. It is shown that this ring is a proper Euclidean domain, and the Euclidean division process is characterized. A topology is defined on this ring, and it is shown that the set of units is open. Certain generalizations of this ring are also discussed.

2.1 THE RING S AS A EUCLIDEAN DOMAIN

Let $\mathbb{R}[s]$ denote the set of polynomials in the indeterminate s with coefficients in the field $\mathbb{R}$ of real numbers. Then $\mathbb{R}[s]$ is a Euclidean domain if the degree of a polynomial is defined in the usual way (see Section A.4). The field of fractions associated with $\mathbb{R}[s]$ is denoted by $\mathbb{R}(s)$, and consists of rational functions in s with real coefficients. One can think of $\mathbb{R}(s)$ as the set of all possible transfer functions of scalar, lumped, linear time-invariant systems.

Now let **S** denote the subset of $\mathbb{R}(s)$ consisting of all rational functions that are bounded at infinity, and whose poles all have negative real parts; i.e., **S** consists of *all proper stable rational functions*. A function $p(\cdot)$ belongs to **S** if and only if it is the transfer function of a BIBO (bounded-input-bounded-output)-stable system. As is customary, let C_+ denote the closed right half-plane $\{s : \operatorname{Re} s \geq 0\}$, and let C_{+e} denote the extended right half-plane, i.e., C_+ together with the point at infinity. Then a rational function belongs to **S** if and only if it has no poles in C_{+e}.

It is left to the reader to verify that, under the usual definitions of addition and multiplication in the field $\mathbb{R}(s)$, the set **S** is a commutative ring with identity, and is a domain (Problem 2.1.1). Moreover, the field of fractions associated with **S** is precisely $\mathbb{R}(s)$. It is clear that the ratio of any two elements $a, b \in$ **S** with $b \neq 0$ belongs to $\mathbb{R}(s)$. To prove the converse, suppose $h \in \mathbb{R}(s)$, and that $h = \alpha/\beta$ where α, β are polynomials. Let n equal the larger of the degrees of α and β, and define

$$f(s) = \frac{\alpha(s)}{(s+1)^n}, \quad g(s) = \frac{\beta(s)}{(s+1)^n} . \qquad (2.1.1)$$

Then $h(s) = f(s)/g(s)$ is a ratio of two elements of **S**.

It is easy to show that a function in **S** is a unit of **S** (i.e., has an inverse in **S**; see Section A.1) if and only if it has no zeros in the extended right half-plane C_{+e}, (Problem 2.1.2). The units of **S** are sometimes referred to as *miniphase functions*. If $f, g \in$ **S**, then f divides g in **S** if and only if every zero of f in C_{+e} is also a zero of g with at least the same multiplicity (Problem 2.1.3). As an illustration of this, let

$$f_1(s) = \frac{s+1}{(s+2)^2}, \ f_2(s) = \frac{s-1}{s+1}, \ f_3(s) = \frac{s-1}{(s+1)^2}, \tag{2.1.2}$$

$$g_1(s) = \frac{s-1}{s+4}, \ g_2(s) = \frac{1}{3(s+3)}, \ g_3(s) = \frac{s-1}{(s+2)^3}. \tag{2.1.3}$$

Then f_1 divides g_2, g_3 but not g_1; f_2 divides g_1, g_3 but not g_2; and f_3 divides g_3 but not g_1, g_2.
We now define a degree function on the ring **S** such that it becomes a Euclidean domain.

Theorem 2.1.1   Define a function $\delta : \mathbf{S} \setminus 0 \to \mathbf{Z}_+$ as follows: If $f \in \mathbf{S} \setminus 0$, then

$$\begin{aligned} \delta(f) &= \text{no. of zeros of } f \text{ in } C_{+e} \\ &= \text{no. of zeros of } f \text{ in } C_+ + \text{ relative degree of } f . \end{aligned} \tag{2.1.4}$$

Then **S** is a proper Euclidean domain with degree function δ.

The proof of Theorem 2.1.1 requires a preliminary result.

Lemma 2.1.2   *Suppose $\alpha, \beta, \omega \in \mathbb{R}[s]$, with α and β coprime. Then there exist $\phi, \psi \in \mathbb{R}[s]$, with the degree of ϕ less than the degree of β, such that*

$$\alpha\phi + \beta\psi = \omega . \tag{2.1.5}$$

Proof. Since α and β are coprime, there exist $\phi_1, \psi_1 \in \mathbb{R}[s]$ such that

$$\alpha\phi_1 + \beta\psi_1 = 1 . \tag{2.1.6}$$

Multiplying both sides of (2.1.6) by ω and denoting $\phi_1\omega$ by ϕ_2, $\psi_1\omega$ by ψ_2 gives

$$\alpha\phi_2 + \beta\psi_2 = \omega . \tag{2.1.7}$$

Now, for *any* $\theta \in \mathbb{R}[s]$, it follows from (2.1.7) that

$$\alpha(\phi_2 - \beta\theta) + \beta(\psi_2 + \alpha\theta) = \omega . \tag{2.1.8}$$

Since $\mathbb{R}[s]$ is a Euclidean domain, by proper choice of θ we can make the degree of $\phi_2 - \beta\theta$ lower than that of β. Choosing such a θ and defining $\phi = \phi_2 - \beta\theta$, $\psi = \psi_2 + \alpha\theta$ gives (2.1.5). $\square$

Proof of Theorem 2.1.1. Let $f, g \in \mathbf{S}$ with $g \neq 0$. It is shown that there exists a q such that $\delta(f - gq) < \delta(g)$, where $\delta(0)$ is taken as $-\infty$. If $\delta(g) = 0$, then g is a unit. In this case, take $q = fg^{-1}$; this gives $f - gq = 0$, $\delta(f - gq) = -\infty < \delta(g)$.

Now suppose $\delta(g) > 0$, and suppose $g = \alpha_g/\beta_g$, where $\alpha_g, \beta_g \in \mathbb{R}[s]$ are coprime. Factor α_g as a product $\alpha_- \gamma$ where all zeros of α_- are in C_- and all zeros of γ are in C_+. Express g in the form

$$g(s) = e(s)\frac{\gamma(s)}{(s+1)^n}, \quad \text{where} \quad e(s) = \frac{\alpha_-(s)(s+1)^n}{\beta_g(s)}, \tag{2.1.9}$$

with $n = \delta(g)$, and observe that e is a unit of $\mathbf{S}$. Let $f = \alpha_f/\beta_f$, where $\alpha_f, \beta_f \in \mathbb{R}[s]$ are coprime. Since all zeros of β_f are in C_- and all zeros of γ are in C_+, it follows that β_f, γ are coprime in $\mathbb{R}[s]$. Hence there exist $\phi, \psi \in \mathbb{R}[s]$ such that the degree of ϕ is less than that of β_f, and

$$\gamma(s)\phi(s) + \beta_f(s)\psi(s) = \alpha_f(s)(s+1)^{n-1}. \tag{2.1.10}$$

Dividing both sides of (2.1.10) by $\beta_f(s)(s+1)^{n-1}$ gives

$$\frac{\gamma(s)}{(s+1)^n} \cdot \frac{\phi(s)(s+1)}{\beta_f(s)} + \frac{\psi(s)}{(s+1)^{n-1}} = \frac{\alpha_f(s)}{\beta_f(s)}. \tag{2.1.11}$$

So if we define

$$q(s) = \frac{\phi(s)(s+1)}{\beta_f(s)} \cdot \frac{1}{e(s)}, \tag{2.1.12}$$

$$r(s) = \frac{\psi(s)}{(s+1)^{n-1}}, \tag{2.1.13}$$

then it follows from (2.1.11) that $f = gq + r$. Moreover, since the degree of ϕ is *less* than that of β_f, q is proper; q is also stable, since all zeros of β_f are in C_-. Hence $q \in \mathbf{S}$. Since $r = f - gq$, r also belongs to $\mathbf{S}$. In particular, r is proper, which shows that the degree of ψ is less than or equal to $n - 1$. Hence, from (2.1.4), $\delta(r) < \delta(g)$. Since the above procedure can be repeated for every $f, g \in \mathbf{S}$ with $g \neq 0$, we have shown that $\mathbf{S}$ is a Euclidean domain with degree function δ.

It is left to the reader to show that $\mathbf{S}$ is not a field and that $\delta(fg) = \delta(f) + \delta(g) \,\forall f, g \in \mathbf{S}$. Hence $\mathbf{S}$ is a *proper* Euclidean domain (see Problem 2.1.4). $\qquad\square$

Let

$$f(s) = \frac{s+4}{s+1}, \quad g(s) = -\frac{2s+3}{s+1}. \tag{2.1.14}$$

Then both f and g are units of $\mathbf{S}$ and accordingly $\delta(f) = \delta(g) = 0$. However,

$$f(s) + g(s) = \frac{-s+1}{s+1} \tag{2.1.15}$$

has a degree of 1. Hence in general it is *not* true that $\delta(f+g) \leq \max\{\delta(f), \delta(g)\}$. As a result, the Euclidean division process does not result in a unique remainder as is the case with polynomials, i.e., given $f, g \in \mathbf{S}$ with $g \neq 0$, there may exist *several* $q \in \mathbf{S}$ such that $\delta(f - gq) < \delta(g)$. For example, let

$$f(s) = \frac{(s-1)(s^2 + 2s + 2)}{(s+1)^3}, \quad g(s) = \frac{s+5}{(s+1)^3}. \qquad (2.1.16)$$

Then $\delta(g) = 2$ since g has no zeros in C_+ but a double zero at infinity. Now suppose c is a nonnegative constant, and consider $\delta(f + cg)$. It is easy to verify using root-locus arguments that

$$\delta(f + cg) = \begin{cases} 1 & \text{if } 0 \leq c \leq 0.4 \\ 0 & \text{if } 0.4 \leq c < 0.5 \\ 2 & \text{if } c \geq 0.5 \end{cases} \qquad (2.1.17)$$

Thus there exist *infinitely many* c such that $\delta(f + cg) < \delta(g)$; moreover the degree of the remainder $\delta(f + cg)$ can be made to assume *several* distinct values less than $\delta(g)$ by suitable choice of c. These phenomena have no analogs in the case of the polynomial ring $\mathbb{R}[s]$. In view of this, an important question is the following: Given $f, g \in \mathbf{S}$ with $g \neq 0$, what is the *smallest* value that $\delta(f - gq)$ can achieve as q varies over $\mathbf{S}$? This question is answered in Section 2.3 (see Theorem 2.3.2).

The primes in the ring $\mathbf{S}$ are functions of the form $1/(s + a)$ where $a > 0$; $(s - a)/(s + b)$ where $a \geq 0, b > 0$; $[(s - a)^2 + b^2]/(s + c)^2$ where $a \geq 0, b, c > 0$; and their associates. Since two functions f and g are coprime if and only if they have no common prime divisors (Fact A.3.8), it follows that f and g are coprime if and only if they have no common zeros in the extended RHP C_{+e}, or equivalently, if they have no common zeros in C_+ and at least one of them has relative degree zero. This leads to the following observation.

Fact 2.1.3   Suppose $f \in \mathbb{R}(s)$ and express f as α/β where $\alpha, \beta \in \mathbb{R}[s]$ have no common zeros in C_+. Let n equal the larger of the degrees of α and β, and define

$$a(s) = \frac{\alpha(s)}{(s+1)^n}, \quad b(s) = \frac{\beta(s)}{(s+1)^n}. \qquad (2.1.18)$$

Then $a, b \in \mathbf{S}$ are coprime, and $f = a/b$ is a reduced form for f viewed as a fraction over $\mathbf{S}$. A point $s \in C_{+e}$, is a zero (resp. pole) of f if and only if it is a zero of a (resp. a zero of b). If so, the order of s as a zero (resp. pole) of f equals its multiplicity as a zero of a (resp. zero of b).

Note that the polynomials α, β defined above *can* have common zeros in the *left* half-plane without violating the above conclusions. As an illustration, suppose

$$f(s) = \frac{(s-1)(s+2)}{(s-2)(s-3)} \qquad (2.1.19)$$

and let

$$\alpha(s) = (s-1)(s+2)(s+4), \quad \beta(s) = (s-2)(s-3)(s+4). \qquad (2.1.20)$$

Then

$$a(s) = \frac{(s-1)(s+2)(s+4)}{(s+1)^3}, \quad b(s) = \frac{(s-2)(s-3)(s+4)}{(s+1)^3} \tag{2.1.21}$$

are coprime in **S**. Note that $s = -4$ is a zero of a even though it is not a zero of f. But this does not contradict Fact 2.1.3, as the latter only applies to points $s \in C_{+e}$.

Quite often, the objective of control system design is not merely to stabilize a given plant but to place the closed-loop poles in some pre-specified region of the left half-plane. In such applications, it is desirable to replace the open left half-plane C_- by a more general *domain of stability* **D** which is open and has the property that $\bar{s} \in \mathbf{D}$ whenever $s \in \mathbf{D}$ (here the bar denotes complex conjugation). In this case, one replaces the set **S** by the set $\mathbf{S_D}$, which consists of all proper rational functions whose poles all lie in **D**. Thus **S** is a special case of $\mathbf{S_D}$, with $\mathbf{D} = C_-$. It is left to the reader to verify that $\mathbf{S_D}$ is a commutative domain with identity; moreover, if we define a function $\delta : \mathbf{S_D} - 0 \to \mathbf{Z}_+$ by

$$\delta(f) = \text{Relative degree of } f + \text{no. of zeros of } f \text{ outside } \mathbf{D} \,, \tag{2.1.22}$$

Then $\mathbf{S_D}$ is a proper Euclidean domain provided the region **D** contains at least one real number (see Problem 2.1.5).

For instance, if it is desired to design continuous-time control systems with prescribed maximum settling time and minimum damping, the domain of stability **D** should be taken as in Figure 2.1.

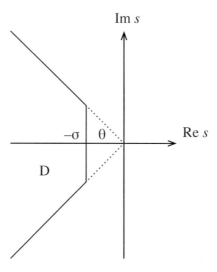

Figure 2.1: Generalized Region of Stability.

In the study of discrete-time control systems, the usual notion of stability is that the unit pulse response of the system is absolutely summable. Thus, if a system has the unit pulse response sequence $\{h_i\}_{i=0}^{\infty}$, the system is BIBO-stable if and only if [26, Chapter 6]

$$\sum_{i=0}^{\infty} |h_i| < \infty . \tag{2.1.23}$$

Define the transfer function of the system by

$$f(z) = \sum_{i=0}^{\infty} h_i z^i . \tag{2.1.24}$$

Then a system is BIBO-stable if and only if the power series defining its transfer function converges absolutely for all z with $|z| \le 1$. If the system is lumped, then the transfer function $f(z)$ is rational. In this case the system is BIBO-stable if and only if $f(\cdot)$ has no poles on or inside the unit circle in the complex plane. The set of such rational functions is denoted by $\mathbf{R}_{\infty}$ (the rationale for this notation can be found in Chapter 6). If we define a function $\delta : \mathbf{R}_{\infty} \setminus 0 \to \mathbf{Z}_+$ by

$$\delta(f) = \text{no. of zeros of } f \text{ inside the closed unit circle,} \tag{2.1.25}$$

then $\mathbf{R}_{\infty}$ becomes a proper Euclidean domain (Problem 2.1.7).

The reader is cautioned that the definition (2.1.24) of the z-transform of a sequence differs from the conventional one, which is

$$g(z) = \sum_{i=0}^{\infty} h_i z^{-i} . \tag{2.1.26}$$

Clearly $g(z) = f(z^{-1})$. The current definition has some advantages, one of which is that every polynomial in $\mathbb{R}[z]$ belongs to $\mathbf{R}_{\infty}$; in fact, elements of $\mathbb{R}[z]$ correspond to so-called finite impulse response (FIR) filters.

There is a close connection between the sets $\mathbf{S}$ and $\mathbf{R}_{\infty}$, which represent the sets of stable transfer functions in the case of continuous-time and discrete-time systems, respectively. Specifically, if we define the bilinear mapping

$$z = \frac{s-1}{s+1} \tag{2.1.27}$$

that takes the extended right half-plane C_{+e} into the closed unit disc $\mathbf{D}$, then there is a one-to-one correspondence between functions in $\mathbf{S}$ and in $\mathbf{R}_{\infty}$ via the association

$$f(s) \mapsto g(z) = f((1+z)/(1-z)) . \tag{2.1.28}$$

The inverse correspondence is of course given by

$$g(z) \mapsto f(s) = g((s-1)/(s+1)) . \tag{2.1.29}$$

PROBLEMS

2.1.1. Show that **S** is a commutative domain with identity.

2.1.2. Show that a function in **S** has an inverse in **S** if and only if it has no zeros in C_{+e}.

2.1.3. Suppose $f, g \in \mathbf{S}$. Show that f divides g in **S** (i.e., $g/f \in \mathbf{S}$) if and only if every zero of f in C_{+e} is also a zero of g of at least the same multiplicity.

2.1.4. Show that **S** is a proper Euclidean domain by showing that $\delta(fg) = \delta(f) + \delta(g) \, \forall f, g \in \mathbf{S} \setminus 0$.

2.1.5. Show that if the region **D** contains at least one real number, then $\mathbf{S_D}$ is a proper Euclidean domain with degree function δ defined in (2.1.22).

2.1.6. Let $\mathbb{R}^-(s)$ consist of those functions in $\mathbb{R}(s)$ whose poles are all in the open left half-plane (but which need not be proper). Show that $\mathbb{R}^-(s)$ is a proper Euclidean domain if we define $\delta(f) = $ no. of zeros of f in C_+.

2.1.7. Show that $\mathbf{R}_\infty$ is a proper Euclidean domain with the degree function δ defined in (2.1.25).

2.2   TOPOLOGY ON S AND M(S)

In this section, we define a topology on the ring **S** of proper stable rational functions, and show that the set **U** of units of **S** is open in the proposed topology. This topology is extended to the set **M(S)** of matrices with elements in **S**, and it is shown that the set **U(S)** of unimodular matrices is open in **M(S)**. The generalizations to the ring $\mathbf{S_D}$ are also indicated.

Definition 2.2.1   The norm function $\| \cdot \| : \mathbf{S} \to \mathbb{R}$ is declined by

$$\| f \| = \sup_{s \in C_+} |f(s)| . \tag{2.2.1}$$

Since every function in **S** is analytic over the open right half-plane, it follows from the maximum modulus theorem [81, p. 229] that

$$\| f \| = \sup_{\omega \in \mathbb{R}} |f(j\omega)| . \tag{2.2.2}$$

The quantity in (2.2.2) can be interpreted simply in terms of the Nyquist plot of $f(\cdot)$: The norm of f is the radius of the smallest circle (centered at the origin) that contains the Nyquist plot of f (see Figure 2.2).

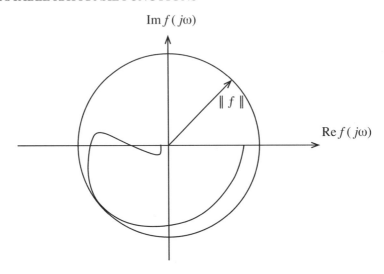

Figure 2.2: Graphical Interpretation of the Norm of a Function.

It can be verified that, with this norm, **S** becomes a normed algebra over the field of real numbers $\mathbb{R}$. Hence there is a natural topology on **S** induced by this norm. For every $f \in \mathbf{S}$ and $\varepsilon > 0$, define the ball $\mathbf{B}(f, \varepsilon)$ by

$$\mathbf{B}(f, \varepsilon) = \{g \in \mathbf{S} : \|f - g\| < \varepsilon\} . \tag{2.2.3}$$

Then the collection of balls $\mathbf{B}(f, \varepsilon)$ forms a base for a topology on **S**. In this topology, a set **A** is open if (and only if), corresponding to every $f \in \mathbf{A}$, there is a ball $\mathbf{B}(f, \varepsilon) \subseteq \mathbf{A}$. A set **B** is closed if and only if its complement is open, or equivalently, the limit of any convergent sequence in **B** again belongs to **B**. Also, a sequence $\{f_i(\cdot)\}$ in **S** converges to $f \in \mathbf{S}$ if and only if $f_i(s) \to f(s)$ uniformly for all $s \in C_+$, or equivalently, if $f_i(j\omega) \to f(j\omega)$ uniformly for all $\omega \in \mathbb{R}$; this follows from (2.2.1) and (2.2.2).

Recall from Section 2.1 that a function $f \in \mathbf{S}$ is a unit of **S** (i.e., $1/f \in \mathbf{S}$) if and only if f has no zeros in the extended right half-plane C_{+e}. An equivalent statement is the following: $f \in \mathbf{S}$ has an inverse in **S** if and only if

$$\inf_{s \in C_+} |f(s)| =: c(f) > 0 . \tag{2.2.4}$$

Further, if f is a unit, then it is immediate that

$$\sup_{s \in C_+} |1/f(s)| = [\inf_{s \in C_+} |f(s)|]^{-1} , \tag{2.2.5}$$

so that $\|f^{-1}\| = 1/c(f)$. In particular, $1 + f$ is a unit whenever $\|f\| < 1$, since

$$c(1 + f) = \inf_{s \in C_+} |1 + f(s)| \tag{2.2.6}$$

$$\geq 1 - \sup_{s \in C_+} |f(s)| = 1 - \|f\| > 0 \,.$$

Moreover, $\|(1 + f)^{-1}\| = 1/c(1 + f) \leq 1/(1 - \|f\|)$. The next lemma presents a generalization of this fact and also establishes one of the most useful properties of the topology on **S**.

Lemma 2.2.2   *Let* **U** *denote the set of units of* **S**, *and suppose* $f \in$ **U**. *Then* $g \in$ **U** *whenever* $\|g - f\| < 1/\|f^{-1}\|$, *in which case*

$$\|g^{-1} - f^{-1}\| \leq \|f^{-1}\| \frac{\|f^{-1}\| \cdot \|g - f\|}{1 - \|f^{-1}\| \cdot \|g - f\|} \,. \tag{2.2.7}$$

As a consequence, **U** *is an open subset of* **S** *and the map* $f \mapsto f^{-1}$ *mapping* **U** *into itself is*

continuous. Finally, the above bound is the best possible in the sense that if f is any unit, then there exists a nonunit g such that $\|g - f\| = 1/\|f^{-1}\|$.

Proof. Suppose $f \in$ **U** and that $\|g - f\| < 1/\|f^{-1}\|$. Then $\|gf^{-1} - 1\| \leq \|g - f\| \cdot \|f^{-1}\| < 1$. It follows from the preceding discussion that $1 + [gf^{-1} - 1] = gf^{-1}$ is a unit, whence g is also a unit. To establish the bound (2.2.7), observe that

$$\|g^{-1} - f^{-1}\| \leq \|g^{-1}\| \cdot \|f - g\| \cdot \|f^{-1}\|$$

$$\leq [\|f^{-1}\| + \|g^{-1} - f^{-1}\|] \cdot \|f - g\| \cdot \|f^{-1}\| \,. \tag{2.2.8}$$

$$[1 - \|f - g\| \cdot \|f^{-1}\|] \cdot \|g^{-1} - f^{-1}\| \leq \|f^{-1}\|^2 \cdot \|f - g\| \,. \tag{2.2.9}$$

Now (2.2.7) follows readily from (2.2.9).

The proof that the bound is the best possible is somewhat involved. Since f is a unit, f^{-1} is analytic over C_+, which implies that f attains its *minimum modulus* over C_+ on the $j\omega$-axis or at infinity. Thus, either $|f(\infty)| = c(f)$ or else there exists a finite ω_0 such that $|f(j\omega_0)| = c(f)$.

Case (i) $|f(\infty)| = c(f)$. Since $f(\infty)$ is real, it equals $\pm c(f)$. Let h equal the *constant* function sign $c(f) \cdot c(f)$. Then $\|h\| = c(f)$ but $g = f - h$ is not a unit since $g(\infty) = (f - h)(\infty) = 0$.

Case (ii) $|f(0)| = c(f)$. Once again $f(0)$ is real and must equal $\pm c(f)$. This is similar to the above case.

Case (iii) $|f(j\omega_0)| = c(f)$ for some finite nonzero ω_0. The additional complication in this case is that $f(j\omega_0)$ could be nonreal. If $f(j\omega_0)$ is real simply choose h as in the above two cases. Otherwise, assume without loss of generality that $\omega_0 > 0$ (since $|f(j\omega_0)| = |f(-j\omega_0)|$), and express $f(j\omega_0)$ in the form $\pm c(f) \exp(j\theta_0)$ where $\theta_0 \in (-\pi, 0]$. Let h be the all-pass function

$$h(s) = \pm c(f) \frac{s - \alpha}{s + \alpha} \tag{2.2.10}$$

where $\alpha \geq 0$ is adjusted so that $h(j\omega_0) = f(j\omega_0)$. Then $\|h\| = c(f)$, but $g = f - h$ is not a unit.

□

Consider the set $\mathbf{M}(\mathbf{S})$ of matrices with elements in the ring $\mathbf{S}$. The norm of a matrix-valued function $A \in \mathbf{M}(\mathbf{S})$ is defined by

$$\|A\| = \sup_{s \in C_+} \bar{\sigma}(A(j\omega)) , \qquad (2.2.11)$$

where $\bar{\sigma}(\cdot)$ denotes the largest singular value of a matrix.[1] Several comments are in order concerning this norm.

(i) Recall that the *Euclidean norm* of a vector x in C^n is defined by

$$\|x\|_2 = (x^*x)^{1/2} . \qquad (2.2.12)$$

where $*$ denotes the conjugate transpose. If $M \in C^{m \times n}$, its *Euclidean induced norm* is defined by

$$\|M\|_2 = \sup_{x \in C^n \setminus 0} \frac{\|Mx\|_2}{\|x\|_2} \qquad (2.2.13)$$

and equals $\bar{\sigma}(M)$. Thus the norm of a matrix $A \in \mathbf{M}(\mathbf{S})$ can also be defined by

$$\|A\| = \sup_{s \in C_+} \|A(s)\|_2 . \qquad (2.2.14)$$

If M, N are complex matrices of compatible dimensions, then $\|MN\|_2 \leq \|M\|_2 \cdot \|N\|_2$. Based on this and (2.2.14), it is immediate if $A, B \in \mathbf{M}(\mathbf{S})$ and the product AB is defined, then

$$\|AB\| \leq \|A\| \cdot \|B\| . \qquad (2.2.15)$$

(ii) In general, even if $A(\cdot)$ is analytic over C_+, the function $s \mapsto \bar{\sigma}(A(s))$ is not. Thus the maximum modulus principle does not apply to $\bar{\sigma}(A(\cdot))$. Nevertheless, it is true that

$$\sup_{\omega \in \mathbb{R}} \bar{\sigma}(A(j\omega)) = \sup_{s \in C_+} \bar{\sigma}(A(s)) = \|A\| . \qquad (2.2.16)$$

Thus the norm of $A \in \mathbf{M}(\mathbf{S})$ can be computed based on the behavior of $A(\cdot)$ on the $j\omega$-axis alone.

The topology on $\mathbf{M}(\mathbf{S})$ is the one induced by the norm (2.2.11). It can be shown to be the same as the product topology on $\mathbf{M}(\mathbf{S})$ obtained from the one on $\mathbf{S}$. In simpler terms, this means the following: Suppose $\{A^l\}$ is a sequence of matrices, and A is another matrix, all of the same order, in $\mathbf{M}(\mathbf{S})$. Then $\|A^l - A\| \to 0$ as $l \to \infty$ if and only if $\|a_{ij}^l - a_{ij}\| \to 0$ as $l \to \infty$ for each i, j. In other words, the sequence of *matrices* $\{A^l\}$ converges to the *matrix* A in the sense of the norm (2.2.11) if and only if each component sequence of *scalars* a_{ij}^l converges to the *scalar* a_{ij} in

---

[1] Recall that the largest singular value of a matrix M is defined as the square-root of the largest eigenvalue of the symmetric matrix M^*M.

the sense of the norm (2.2.1). Thus convergence in $\mathbf{M}(\mathbf{S})$ is equivalent to component-wise uniform convergence over C_+ (or over the $j\omega$-axis).[2]

The next result generalizes Lemma 2.2.2 to the matrix case.

Lemma 2.2.3   *Suppose F is a unit of $\mathbf{M}(\mathbf{S})$ and that $\|G - F\| < \|F^{-1}\|$. Then G is also a unit, and*

$$\|G^{-1} - F^{-1}\| \le \|F^{-1}\| \frac{\|F^{-1}\| \cdot \|G - F\|}{1 - \|F^{-1}\| \cdot \|G - F\|} . \tag{2.2.17}$$

Moreover, this bound is the best possible in the sense that if F is any unit of $\mathbf{M}(\mathbf{S})$, then there exists a nonunit matrix G such that $\|G - F\| = \|F^{-1}\|$.

Proof. Recall that G is a unit of $\mathbf{M}(\mathbf{S})$ if and only if $|G|$ is a unit of $\mathbf{S}$. We begin by showing that if $R \in \mathbf{M}(\mathbf{S})$ and $\|R\| < 1$, then $I + R$ is a unit of $\mathbf{M}(\mathbf{S})$. For each $s \in C_{+e}$, the spectrum of the matrix $I + R(s)$ consists of $\{1 + \lambda_i(s), i = 1, \cdots, n\}$, where $\lambda_1(s), \cdots, \lambda_n(s)$ are the eigenvalues of $R(s)$. Since $\|R\| < 1$, it follows that the norm of the *matrix* $R(s)$ is less than one for all $s \in C_{+e}$. Hence $|\lambda_i(s)| \le \|R(s)\| < 1 \, \forall i, \, \forall s \in C_{+e}$. As a result $1 + \lambda_i(s) \ne 0 \, \forall i, \, \forall s \in C_{+e}$, whence $|I + R(s)| = \prod_i (1 + \lambda_i(s)) \ne 0 \, \forall s \in C_{+e}$. This shows that $|I + R|$ is a unit of $\mathbf{S}$, so that $I + R$ is a unit of the matrix ring $\mathbf{M}(\mathbf{S})$.

Now if F is a unit and $\|G - F\| < 1/\|F^{-1}\|$, then $\|GF^{-1} - I\| < \|G - F\| \cdot \|F^{-1}\| < 1$, so that $GF^{-1} = I + (GF^{-1} - I)$ is a unit of $\mathbf{M}(\mathbf{S})$, by the preceding argument. Thus G is also a unit.

The proof of the inequality (2.2.17) is entirely analogous to that of (2.2.7) and is left to the reader.

To prove that the bound is the best possible, we construct a matrix $H \in \mathbf{M}(\mathbf{S})$ such that $\|H\| = \|F^{-1}\|$ but $G = F - H$ is not a unit. The argument exactly follows that in the scalar case after taking into account some additional complications arising due to the fact of dealing with matrices. Now (2.2.14) shows that either $\|F^{-1}(\infty)\|_2 = \|F^{-1}\|$, or else $\|F^{-1}(j\omega)\|_2 = \|F^{-1}\|$ for some finite ω.

Case (i) $\|F^{-1}(\infty)\|_2 = \|F^{-1}\|$. Since $F^{-1}(\infty)$ is a *real* matrix, there exists a *real* vector v such that $\|v\|_2 = 1$, and $\|F^{-1}(\infty)v\|_2 = \|F^{-1}(\infty)\|_2$. For brevity let c denote the constant $\|F^{-1}(\infty)\|_2 = \|F^{-1}\|$, and let u denote the vector $F^{-1}(\infty)v$. Now let H equal the constant matrix vu'/c^2. Then routine calculations show that $\|H\| = 1/c$. However, $[F(\infty) - H]u = 0$, which shows that $|(F - H)(\infty)| = 0$. Thus $G = F - H$ is not a unit.

Case (ii) $\|F^{-1}(0)\|_2 = \|F^{-1}\|$. Since $F^{-1}(0)$ is a real matrix, the above proof applies.

Case (iii) $\|F^{-1}(j\omega)\|_2 = \|F^{-1}\|$ for some finite nonzero ω, which can be assumed to be positive without loss of generality. As in the scalar case, the possible complication is that $F(j\omega)$ may

---

[2]The reason for drawing attention to this point at great length is that this is a property special to *stable* systems, and is not true for sequences of *unstable* transfer matrices; see Section 7.2.

not be a real matrix. Select a complex vector v such that $\|v\|_2 = 1$ and $\|F^{-1}(j\omega)v\|_2 = \|F^{-1}\|$. As before let c denote $\|F^{-1}\|$ and let u denote $F^{-1}(j\omega)v$. Express each component of u, v in the form

$$u_i = \bar{u}_i \exp(j\theta_i), \quad v_i = \bar{v}_i \exp(j\phi_i), \quad i = 1, \cdots, n . \tag{2.2.18}$$

where $\bar{u}_i, \bar{v}_i$ are real and $\theta_i, \phi_i \in (-\pi, 0] \forall i$. Define the vector-valued functions $a(s), b(s)$ by

$$a_i(s) = \bar{v}_i \frac{s - \alpha_i}{s + \alpha_i}, \quad b_i(s) = \bar{u}_i \frac{s - \beta_i}{s + \beta_i} , \tag{2.2.19}$$

where the nonnegative constants α_i, β_i are chosen such that $a(j\omega) = v, b(j\omega) = u$. Finally, define $H(s) = c^{-2}a(s)b'(s)$. Then $\|H\| = c$, but $(F - H)(j\omega)$ is singular, so that $G = F - H$ is not a unit. □

All of the preceding development can be carried out in the more general ring $\mathbf{S_D}$, consisting of all proper rational functions whose poles all lie in the domain of stability $\mathbf{D}$, subject to a few technical conditions. Specifically, the norm of a function f in $\mathbf{S_D}$ is defined as

$$\|f\| = \sup_{s \in \mathbf{D}^c} |f(s)| , \tag{2.2.20}$$

where the superscript "c" denotes the complement of a set. In order for this norm to be well-defined, it is necessary for the set $\mathbf{D}$ to be open, or equivalently, for its complement to be closed.

Finally, consider the case of discrete-time systems, where one is interested in the set $\mathbf{R}_\infty$, consisting of all rational functions that are analytic on the closed unit disc. In this case one can define the norm of a function f in $\mathbf{R}_\infty$ as

$$\|f\| = \sup_{|z| \leq 1} |f(z)| . \tag{2.2.21}$$

All of the previously derived bounds etc. carry over to either of these more general situations.

In conclusion, note that the correspondence between the sets $\mathbf{S}$ and $\mathbf{R}_\infty$ defined in (2.1.27–2.1.28) is actually norm-preserving if the norms on these two algebras are defined as in (2.2.1) and (2.2.21), respectively.

2.3   EUCLIDEAN DIVISION IN S

As shown in Section 2.1, the ring $\mathbf{S}$ is a proper Euclidean domain if the degree of a function f is defined as the number of zeros of f in the extended right half-plane C_{+e}. Thus, given $f, g \in \mathbf{S}$ with $g \neq 0$, there exists a $q \in \mathbf{S}$ such that $\delta(f - gq) < \delta(g)$. However, such a q need not to be unique. The main objective of this section is to answer the question: What is the smallest value that $\delta(f - gq)$ can achieve as q is varied? As a prelude to answering this question, we derive a very useful interpolation result. It turns out that this result can be proved in two ways: one proof is elementary but very tedious, and this is the one given in this section. A much more elegant proof requiring more advanced concepts is given in the next section.

Recall that a function f in $\mathbf{S}$ is a unit if and only if f has no zeros in C_{+e}, or equivalently $\delta(f) = 0$. Now suppose $S = \{s_1, \cdots, s_n\}$ is a set of points in C_{+e}, $M = \{m_1, \cdots, m_n\}$ is a corresponding set of positive integers, and $R = \{r_{ij}, j = 0, \cdots, m_i - 1, i = 1, \cdots, n\}$ is a corresponding set of complex numbers. We are interested in knowing whether or not there exists a unit $u(\cdot)$ in $\mathbf{S}$ that satisfies the interpolation constraints

$$\frac{d^j}{ds^j}u(s_i) = r_{ij}, \quad j = 0, \cdots, m_i - 1; \quad i = 1, \cdots, n, \tag{2.3.1}$$

where the zeroth order derivative of u is taken as u itself. In other words, we would like to know whether or not there exists a unit u a such that its functional and derivative values at specified points s_i in C_{+e} are equal to the specified values r_{ij}.

Before tackling the problem, a few simplifying observations are made. First, since u is a rational function with *real* coefficients, as are all its derivatives,

$$\frac{d^j}{ds^j}u(s_0) = \overline{\left\{ \frac{d^j}{ds^j}u(\bar{s}_0) \right\}} \,\forall s_0 \in C_{+e}, \tag{2.3.2}$$

where the bar denotes complex conjugation. Thus, in order for a unit u to exist that satisfies (2.3.1), two necessary conditions must be satisfied: (i) If s_i is real for some i, then the corresponding r_{ij} must be real for all j. (ii) If $s_i = \bar{s}_k$ for some i, k, then $r_{ij} = \bar{r}_{kj}$ for all j. Thus we may as well assume at the outset that Im $s_i \geq 0$ for all i, and that r_{ij} is real whenever s_i is real. There is one other obvious necessary condition: If u is a unit, it can have no zeros in C_{+e}. Hence, the *function value* $r_{i0} \neq 0$ for all i. However, the constants r_{ij}, $j \geq 1$, which are the prescribed values of the derivatives of u, may be zero for some i and j.

Theorem 2.3.1 below gives an extremely simple necessary and sufficient condition for the existence of a unit u satisfying (2.3.1). However, the proof is anything but simple.

Theorem 2.3.1   Let $\sigma_1, \cdots, \sigma_l$ be distinct nonnegative extended real numbers,[3] and let $s_{l+1}, \cdots, s_n$ be distinct complex numbers with positive imaginary part. Let $S = \{\sigma_1, \cdots, \sigma_l, s_{l+1}, \cdots, s_n\}$, let $M = \{m_1, \cdots, m_n\}$ be a corresponding set of positive integers, and let $R = \{r_{ij}, j = 0, \cdots, m_i - 1, i = 1, \cdots, n\}$ be a set of complex numbers with r_{ij} real whenever $j = 0, \cdots, m_i - 1, i = 1, \cdots, l$, and $r_{i0} \neq 0$ for all i. Under these conditions, there exists a unit u in $\mathbf{S}$ satisfying (2.3.1) if and only if the numbers $r_{10}, \cdots, r_{l0}$ are all of the same sign.

The proof of Theorem 2.3.1 is given at the end of the section, as it is long and tedious. The next step is to study Euclidean division in the ring $\mathbf{S}$.

Suppose $f, g \in \mathbf{S}$ are given, with $g \neq 0$. With the aid of Theorem 2.3.1 it is possible to compute the minimum value attained by $\delta(f - gq)$ as q varies over $\mathbf{S}$. For this purpose, we may as well assume at the outset that f and g are coprime. To see why this is so, suppose f and g are not

---

[3]This implies that at most one of the σ_i can be infinite.

coprime, and let $f = f_1 d$, $g = g_1 d$, where d is a greatest common divisor of f and g. Then, since **S** is a proper Euclidean domain,

$$\delta(f - gq) = \delta(d) + \delta(f_1 - qg_1) . \tag{2.3.3}$$

Thus the problem of minimizing $\delta(f - gq)$ is equivalent to minimizing $\delta(f_1 - qg_1)$, where f_1 and g_1 are coprime.

Theorem 2.3.2   Suppose $f, g \in$ **S** are coprime, with $g \neq 0$, and let $\sigma_1, \cdots, \sigma_l$ be the real nonnegative zeros of $g(\cdot)$ in C_{+e} (if any), arranged in ascending order, with $\sigma_l = \infty$ if appropriate. Let v equal the number of sign changes in the sequence $\{f(\sigma_1), \cdots, f(\sigma_l)\}$.[4] Then

$$\min_q \delta(f - gq) = v . \tag{2.3.4}$$

Proof. The proof is divided into two parts: In the first part, it is shown that $\delta(f - gq) \geq v$ for all q. In the second part, it is shown that this lower bound is exact.

To show that $\delta(f - gq) \geq v$ for all q, note that

$$(f - gq)(\sigma_i) = f(\sigma_i) \quad \text{for} \quad i = 1, \cdots, l . \tag{2.3.5}$$

Therefore the sequence $\{(f - gq)(\sigma_1), \cdots, (f - gq)(\sigma_l)\}$ contains v sign changes, whatever be q. Hence the function $f - gq$ has at least v real zeros in C_{+e}. This shows that $\delta(f - gq) \geq v$ for all q.

Next it is shown that there exists a q such that $\delta(f - gq) = v$. Towards this end, select $a, b \in$ **S** such that

$$af + bg = 1 . \tag{2.3.6}$$

and note that a and g are also coprime as a consequence. Now (2.3.6) shows that $(af)(\sigma_i) = 1$ for all i, which shows that $a(\sigma_i)$ and $f(\sigma_i)$ have the same sign for all i. Now select a function $h \in$ **S** such that $\delta(h) = v$ and such that the sequence $\{(ah)(\sigma_1), \cdots, (ah)(\sigma_l)\}$ has no sign changes.[5] Let **S** denote the set of C_{+e}-zeros of the function g. (This set of course includes the points $\sigma_1, \cdots, \sigma_l$). By construction, the function ah has the same sign at all real C_{+e}-zeros of the function g. Hence, by Theorem 2.3.1, there exists a unit, say u, that interpolates the values of ah and its derivatives at the C_{+e}-zeros of g, or equivalently, such that $u - ah$ is divisible by g in **S**. Suppose $u - ah = cg$, and rewrite this in the form

$$ah + cg = u . \tag{2.3.7}$$

---

[4]Note that $f(\sigma_i) \neq 0$ for all i, since f and g are coprime; hence the sign of $f(\sigma_i)$ is well-defined.
[5]This can be accomplished by choosing the function h to have a simple real zero lying between σ_i, and σ_{i+1} whenever $a(\sigma_i)$ and $a(\sigma_{i+1})$ have different signs.

Dividing both sides of (2.3.7) by u gives

$$ar + pg = 1 , \tag{2.3.8}$$

where $r = h/u$, $p = c/u$. Observe also that

$$
\begin{aligned}
\delta(r) &= \delta(h/u) \\
&= \delta(h) + \delta(1/u) \\
&= \delta(h) \text{ since } 1/u \text{ is a unit} \\
&= v \text{ by construction .}
\end{aligned}
\tag{2.3.9}
$$

The proof is concluded by showing that $r = f - gq$ for some $q \in \mathbf{S}$, or equivalently, that g divides $f - r$. Subtracting (2.3.8) from (2.3.6) gives

$$a(f - r) = g(p - b) . \tag{2.3.10}$$

Thus g divides the product $a(f - r)$. Since a and g are coprime, it must be that g divides $f - r$ (see Lemma A.11). □

This section is concluded by giving a proof of Theorem 2.3.1.

Proof of Theorem 2.3.1. "only if" Suppose $r_{i0} > 0$ and $r_{k0} < 0$ for some $i, k \in \{1, \cdots, l\}$. Then *any* function satisfying (2.3.1) must have a real zero between σ_i and σ_k and therefore cannot be a unit.

"if" This part of the proof is constructive.[6] We show that, if $r_{10}, \cdots, r_{l0}$ are all of the same sign, then it is possible to construct a unit u satisfying (2.3.1) after $m = m_1 + \cdots + m_n$ repetitions of the following algorithm. Because of the extremely involved nature of the proof, it is divided into four parts.

First, suppose we have found a unit u such that

$$u^{(j)}(\sigma_i) = r_{ij}, \ j = 0, \cdots, m_i - 1; \ i = 1, \cdots, k - 1, k \leq l . \tag{2.3.11}$$

We show how to construct a unit v such that v satisfies (2.3.11) (with u replaced by v) and in addition, satisfies $v(\sigma_k) = r_{k0}$. In other words, v interpolates all the same values as u, plus one extra function value at σ_k, which happens to be a real number. If $k = 1$ so that the condition (2.3.11) is vacuous, simply choose $v = r_{k0}$. Otherwise, proceed as follows: Define

$$f(s) = \prod_{i=1}^{k-1} \left[\frac{s - \sigma_i}{s + 1} \right]^{m_i} . \tag{2.3.12}$$

Then $f \in \mathbf{S}$; moreover, $1 + bf$ is a unit whenever $|b| < 1/\|f\|$. For such a b, the function $(1 + bf)^a$ is also a unit for all positive integers a. We shall show how to select b and a such that $|b| < 1/\|f\|$ and such that

$$v = (1 + bf)^a u \tag{2.3.13}$$

[6]The reader is reminded that Section 2.4 contains a simple but advanced proof of Theorem 2.3.1.

satisfies $v(\sigma_k) = r_{k0}$ as well as (2.3.11).

Using the binomial expansion,

$$v = u + u \sum_{j=1}^{a} q_{aj}(bf)^j , \qquad (2.3.14)$$

where $q_{aj} = a!/(a-j)!j!$ is the binomial coefficient. It is immediate from (2.3.12) that f and its first $m_i - 1$ derivatives vanish at σ_i, for all $i \leq k - 1$. The same is true of all higher powers of f and thus of uf^j for all j. Hence v continues to interpolate the same function and derivative values as u; that is,

$$v^{(j)}(\sigma_i) = r_{ij}, j = 0, \cdots, m_i - 1; i = 1, \cdots, k - 1 . \qquad (2.3.15)$$

Next,

$$v(\sigma_k) = [1 + bf(\sigma_k)]^a u(\sigma_k) . \qquad (2.3.16)$$

It is routine to compute from (2.3.16) that, to achieve $v(\sigma_k) = r_{k0}$, one must have

$$b = \frac{1}{f(\sigma_k)} \left\{ [r_{k0}/u(\sigma_k)]^{1/a} - 1 \right\} . \qquad (2.3.17)$$

Note that $f(\sigma_k) \neq 0$ from (2.3.12). Also, since u is a unit and satisfies (2.3.11), both $u(\sigma_k)$ and r_{k0} have the same sign, so that the fractional power in (2.3.17) is well-defined. Now, the right side of (2.3.17) approaches zero as $a \to \infty$. Hence, for large enough a, b satisfies $|b| < 1/\|f\|$. For any such b and a, the function $(1 + bf)^a$ is a unit, so that v given by (2.3.13) is a unit and also satisfies $v(\sigma_k) = r_{k0}$.

The remaining steps in the proof follow the same general lines. In the second step, suppose we are still working on the interpolation conditions involving the real values σ_i, and suppose we have found a unit u that satisfies

$$u^{(j)}(\sigma_i) = r_{ij}, j = 0, \cdots, m_i - 1; i = 1, \cdots, k - 1, k \leq l , \qquad (2.3.18)$$
$$u^{(j)}(\sigma_k) = r_{kj}, j = 0, \cdots, t - 1, t \leq m_k . \qquad (2.3.19)$$

In other words, u fully meets the interpolation conditions at the first $k - 1$ points and partially meets them at the k-th point. We shall show how to construct a unit that meets the next interpolation constraint, namely the next derivative constraint at σ_k.

Define

$$f(s) = \left\{ \frac{s - \sigma_k}{s + 1} \right\}^t \prod_{i=1}^{k-1} \left\{ \frac{s - \sigma_i}{s + 1} \right\}^{m_i} . \qquad (2.3.20)$$

Then $f \in \mathbf{S}$. Moreover, all derivatives of f up to order $m_i - 1$ vanish at σ_i, $1 \leq i \leq k - 1$, and all derivatives of f up to order $t - 1$ vanish at σ_k. Let

$$v = (1 + bf)^a u . \qquad (2.3.21)$$

As before, it follows from (2.3.14) that v satisfies (2.3.18)–(2.3.19). Now expand (2.3.14) further as

$$v = u + aubf + u \sum_{j=2}^{a} q_{aj}(bf)^j \ . \tag{2.3.22}$$

Then

$$v^{(t)}(\sigma_k) = u^{(t)}(\sigma_k) + (aubf)^{(t)}(\sigma_k) \ , \tag{2.3.23}$$

because higher powers of bf have a zero of multiplicity greater than t at σ_k. Now, by Leibnitz' rule,

$$\begin{aligned}(aubf)^{(t)}(\sigma_k) &= ab \sum_{j=0}^{t} q_{aj} u^{(t-j)}(\sigma_k) f^{(j)}(\sigma_k) \\ &= ab\, u(\sigma_k) f^{(t)}(\sigma_k) \ . \end{aligned} \tag{2.3.24}$$

Hence, if[7]

$$b = \frac{r_{kt} - u^{(t)}(\sigma_k)}{af^{(t)}(\sigma_k)u(\sigma_k)} \ , \tag{2.3.25}$$

then $v^{(t)}(\sigma_k) = r_{kt}$. Finally, if a is sufficiently large, then $|b| < \|f\|^{-1}$, so that $1 + bf$, $(1 + bf)^a$, and v are all units.

Third, suppose we have found a unit a such that

$$\begin{aligned} u^{(j)}(\sigma_i) &= r_{ij}, \ j = 0, \cdots, m_{i-1}; i = 1, \cdots, l \ , \\ u^{(j)}(s_i) &= r_{ij}, \ j = 0, \cdots, m_{i-1}; i = l+1, \cdots, k-1 \ . \end{aligned} \tag{2.3.26}$$

where $k \leq n$. We now show how to find a unit v such that v satisfies (2.3.25) and also $v(s_k) = r_{k0}$.
Define

$$f(s) = \frac{s - \beta}{s + \beta} \prod_{i=1}^{l} \left[\frac{s - \sigma_i}{s + 1} \right]^{m_i} \prod_{i=l+1}^{k-1} \left[\frac{(s - s_i)(s - \bar{s}_i)}{(s + 1)^2} \right]^{m_i} \tag{2.3.27}$$

$$v = (1 + bf)^a u \ , \tag{2.3.28}$$

where the constants $\beta > 0$, b (real) and a (integer) are yet to be chosen. As before v satisfies (2.3.26), and it only remains to select these constants such that v is a unit and $v(s_k) = r_{k0}$.
Since Im $s_i > 0$, the quantity

$$q := \inf_{\beta \geq 0} \left| \frac{s_k - \beta}{s_k + \beta} \right| \tag{2.3.29}$$

is well-defined and positive. Let $f_1(s) = (s - \beta)/(s + \beta)$ and define $f_2(s) = f(s)/f_1(s)$. Thus $f_2(s)$ is the product of all terms on the right side of (2.3.27) except $(s - \beta)/(s + \beta)$. Now choose

---

[7]Note that $f^{(t)}(\sigma_k) \neq 0$ from (2.3.20) and $u(\sigma_k) \neq 0$ because u is a unit.

the integer a sufficiently large that

$$\left| \left\{ \frac{r_{k0}}{u(s_k)} \right\}^{1/a} - 1 \right| \|f_2\| < q|f_2(s_k)| , \tag{2.3.30}$$

where $(\cdot)^{1/a}$ denotes the a-th principal root. Note that such an a can always be found because the left side of (2.3.30) approaches zero as a approaches infinity. Define

$$b = \frac{[r_{k0}/u(s_k)]^{1/a} - 1}{f_1(s_k)f_2(s_k)} , \tag{2.3.31}$$

where by varying the constant β the argument of $f_1(s_k)$ can be adjusted so as to make b a real number. Since f_1 is an "all-pass" function, $|f_1(j\omega)| = 1$ for all ω, and as a result $\|f\| = \|f_2\|$. The inequalities (2.3.29) and (2.3.30) thus imply that $\|bf\| < 1$. Therefore, $1 + bf$, $(1 + bf)^a$, and v are all units, and moreover $v(s_k) = r_{k0}$ from (2.3.31).

Fourth and last, suppose we have found a unit u such that

$$u^{(j)}(s_i) = r_{ij}, \; j = 0, \cdots, m_i - 1; i = 1, \cdots, k - 1 , \tag{2.3.32}$$
$$u^{(j)}(s_k) = r_{kj}, \; j = 0, \cdots, t - 1 . \tag{2.3.33}$$

where $l < k \le n, t < m_k$ and we have used s_i as a shorthand for σ_i when $i = 1, \cdots, l$. We will find a unit v that satisfies (2.3.32)–(2.3.33) and in addition satisfies $v^{(t)}(s_k) = r_{kt}$.

Define

$$f(s) = \frac{s - \beta}{s + \beta} \prod_{i=1}^{l} \left[\frac{s - \sigma_i}{s + 1} \right]^{m_i} \prod_{i=l+1}^{k-1} \left[\frac{(s - s_i)(s - \bar{s}_i)}{(s + 1)^2} \right]^{m_i}$$
$$\cdot \left[\frac{(s - s_k)(s - \bar{s}_k)}{(s + 1)^2} \right]^{t} . \tag{2.3.34}$$
$$v = (1 + bf)^a u . \tag{2.3.35}$$

Then, as before, v satisfies (2.3.32)–(2.3.33). Using the same reasoning as in (2.3.22)–(2.3.24), it follows that

$$v^{(t)}(s_k) = u^{(t)}(s_k) + ab\, u(s_k) f^{(t)}(s_k) . \tag{2.3.36}$$

As before, let $f(s) = f_1(s)f_2(s)$, where $f_1(s) = (s - \beta)/(s + \beta)$. By Leibnitz' rule,

$$f^{(t)}(s_k) = \sum_{i=0}^{t} q_{ti} f_1^{(t-i)}(s_k) f_2^{(i)}(s_k)$$
$$= f_1(s_k) f_2^{(t)}(s_k) , \tag{2.3.37}$$

since $f_2^{(i)}(s_k) = 0$ for $i < t$. Substitution of (2.3.37) into (2.3.36) gives

$$v^{(t)}(s_k) = u^{(t)}(s_k) + ab\, u(s_k) f_1(s_k) f_2^{(t)}(s_k) . \tag{2.3.38}$$

Accordingly, $v^{(t)}(s_k) = r_{kt}$ if we choose

$$b = \frac{r_{kt} - u^{(t)}(s_k)}{a\, u(s_k)\, f_1(s_k)}. \tag{2.3.39}$$

By selecting a sufficiently large and by adjusting β, we can ensure that b is real and has magnitude less than $\| f \|$. For such a choice of b, v is given by (2.3.35). $\qquad\square$

2.4 INTERPOLATION IN THE DISC ALGEBRA

The main objective of this section is to give a very simple proof of Theorem 2.3.1. Though the proof is very simple in detail, it uses a few advanced concepts, specifically that of a logarithm of an element of a Banach algebra.

Recall that a pair $(\mathbf{B}, \| \cdot \|)$ is a *Banach algebra* if

(i) $(\mathbf{B}, \| \cdot \|)$ is a Banach space,

(ii) $\mathbf{B}$ is an associative algebra over the real or complex field, and

(iii) $\|ab\| \leq \|a\| \cdot \|b\|$ for all $a, b \in \mathbf{B}$.

$\mathbf{B}$ is *commutative* if $ab = ba$ for all $a, b \in \mathbf{B}$, and *has an identity* if there is an element $1 \in \mathbf{B}$ such that $1 \cdot a = a \cdot 1 = a$ for all $a \in \mathbf{B}$. An element $a \in \mathbf{B}$ is a *unit* of $\mathbf{B}$ if there exists a $b \in \mathbf{B}$ such that $ab = ba = 1$ (assuming of course that $\mathbf{B}$ has an identity).

Now suppose that $\mathbf{B}$ is a commutative Banach algebra with identity, and let $\mathbf{U}$ denote the set of its units. This set is nonempty since $1 \in \mathbf{U}$. For each $f \in \mathbf{B}$, the element

$$\exp(f) = \sum_{i=0}^{\infty} f^i / i! \tag{2.4.1}$$

is well-defined. An element $f \in \mathbf{B}$ is said to *have a logarithm* if there exists a $g \in \mathbf{B}$ such that $f = \exp(g)$. If f has a logarithm g, then $f \cdot \exp(-g) = 1$, so that f must necessarily be a unit of $\mathbf{B}$. Thus only units can have logarithms. This naturally leads one to ask whether *every* unit has a logarithm.

Lemma 2.4.1 *A unit $f \in \mathbf{B}$ has a logarithm if and only if f is homotopic to the identity in $\mathbf{U}$, i.e., if and only if there exists a continuous function $h : [0, 1] \to \mathbf{U}$ such that $h(0) = 1, h(1) = f$.*

Proof. "only if" Suppose f has a logarithm, say g. Define $h(\lambda) = \exp(\lambda g)$. Then $h(\cdot)$ is a continuous map from $[0, 1]$ into $\mathbf{U}$ such that $h(0) = 1, h(1) = f$.

"if" This part of the proof proceeds in three steps.

Step 1 If $x \in \mathbf{B}$ and $\|x\| < 1$, then $1 - x$ has a logarithm. In fact the standard Taylor series expansion

$$\sum_{i=1}^{\infty} (-1)^{i-1} x^i / i =: z \tag{2.4.2}$$

converges whenever $\|x\| < 1$, and it is routine to verify that $\exp(z) = 1 - x$.

Step 2 If x has a logarithm and $\|y - x\| < 1/\|x\|^{-1}$, then y has a logarithm. To show this, note that

$$y x^{-1} = [x - (x - y)] x^{-1} = 1 - (x - y) x^{-1} \tag{2.4.3}$$

has a logarithm since $\|(x - y) x^{-1}\| < 1$. If $x = \exp(a)$ and $y x^{-1} = \exp(b)$, then since $\mathbf{B}$ is commutative, it follows that $y = (y x^{-1}) x = \exp(a + b)$.

Step 3 Suppose $h = [0, 1] \to \mathbf{U}$ is continuous and satisfies $h(0) = 1, h(1) = f$. Since the function $u \mapsto u^{-1}$ is continuous, it follows that the function $\lambda \mapsto [h(\lambda)]^{-1}$ mapping $\mathbf{U}$ into itself is also continuous. Moreover, since $[0, 1]$ is compact, this function is also bounded on $[0, 1]$. Accordingly, suppose r is a finite real number such that $\|[h(\lambda)]^{-1}\| < r$ for all $\lambda \in [0, 1]$. Using once again the compactness of $[0, 1]$, we see that the function h is *uniformly* continuous on $[0, 1]$. Thus there exists a $\delta > 0$ such that $\|h(\lambda) - h(\mu)\| < 1/r$ whenever $\lambda, \mu \in [0, 1]$ and $|\lambda - \mu| < \delta$. Let m be an integer larger than $1/\delta$, and let $x_i = h(i\lambda/m)$. Then $x_0 = 1$, and $x_m = f$. Moreover, $\|x_i^{-1}\| \leq r$ and $\|x_{i+1} - x_i\| < 1/r$. From Step 2, we see that since x_0 has a logarithm, so does x_1; by induction, x_i has a logarithm for all i, and in particular f has a logarithm. $\square$

Lemma 2.4.1 is false if $\mathbf{B}$ is noncommutative. In particular, Step 2 of the "if" part of the proof breaks down in the noncommutative case since the product of two units, each of which has a logarithm, need not itself have a logarithm (since $\exp(a + b) \neq \exp(a) \exp(b)$ in general). If a unit $f \in \mathbf{B}$ is homotopic to the identity, then all one can say is that f is a finite product of units, each of which has a logarithm. In other words, f belongs to the group generated by the range of the "exp" function if and only if f is homotopic to the identity in $\mathbf{U}$. This statement can be proved by slightly modifying the proof of Lemma 2.4.1.

Now we introduce the *disc algebra A*, which plays an important role in the control theory of linear distributed systems. The set A consists of all continuous functions mapping the closed unit disc $\mathbf{D}$ into the complex numbers which have the additional property that they are analytic on the interior of $\mathbf{D}$ (i.e., the open unit disc). If addition and multiplication of two functions in A are defined pointwise, then A becomes a commutative Banach algebra with identity over the complex field. Now let A_s denote the subset of A consisting of all symmetric functions; i.e.,

$$A_s = \{f \in A : \bar{f}(\bar{z}) = f(z) \, \forall z \in \mathbf{D}\} \tag{2.4.4}$$

where the bar denotes complex conjugation. Then, A_s is a commutative Banach algebra with identity over the *real* field. The connection between A_s and the set $\mathbf{S}$ of proper stable rational functions is recalled from Section 2.1: Given $f \in \mathbb{R}(s)$, define

$$g(z) = f((s - 1)/(s + 1)) . \tag{2.4.5}$$

Since the bilinear transformation $z = (s - 1)/(s + 1)$ maps the extended right half-plane C_{+e} onto the unit disc $\mathbf{D}$, we see that $g(\cdot)$ is a rational function belonging to A_s if and only if $f \in \mathbf{S}$. Thus if we define

$$\mathbf{R}_\infty = \{ f \in A_s : f \text{ is rational} \} , \tag{2.4.6}$$

then $\mathbf{R}_\infty$ is isometrically isomorphic to $\mathbf{S}$. However, while $\mathbf{R}_\infty$ is a normed algebra, it is not complete; its completion is A_s. Thus every element in A_s can be approximated arbitrarily closely by a polynomial or a rational function in A_s.

In the discrete-time case, a lumped system is BIBO-stable if and only if its transfer function belongs to $\mathbf{R}_\infty$. Finally, note that every polynomial is in A_s.

It is easy to see that $f \in A_s$ is a unit of A_s, if and only if $f(z) \neq 0$ for all $z \in \mathbf{D}$. The next lemma specifies which units in A_s have logarithms.

Lemma 2.4.2   *A unit f of A_s has a logarithm if and only if $f(z) > 0$ for all $z \in [-1, 1]$.*

Proof. "only if" Suppose $f = \exp(g)$ where $g \in A_s$. Since $\bar{g}(\bar{z}) = g(z) \; \forall z \in \mathbf{D}$, it follows that $g(z)$ is real whenever $z \in [-1, 1]$. Hence $f(z) = \exp(g(z)) > 0$ whenever $z \in [-1, 1]$.

"if" Suppose $f(z) > 0$ for all $z \in [-1, 1]$. It is shown that f is homotopic to the identity in the set $\mathbf{U}$ of units of A_s; it will then follow from Lemma 2.4.1 that f has a logarithm. Define

$$h_1(z, \lambda) = f(\lambda z) \quad \text{for} \quad \lambda \in [0, 1] . \tag{2.4.7}$$

Then $h_1(\cdot, 1) = f(\cdot)$ and $h_1(\cdot, 0)$ is the *constant* function $f(0)$. Clearly $h_1(\cdot, \lambda)$ is a continuous function of λ. Moreover, since f is a unit of A_s, we have $f(z) \neq 0$ whenever $|z| \leq 1$. Consequently $f(z) \neq 0$ whenever $|z| \leq \lambda \in [0, 1]$, i.e., $h_1(\cdot, \lambda)$ is a unit for all $\lambda \in [0, 1]$. Therefore, f is homotopic to the constant function $f(0)$ in $\mathbf{U}$. Next, since $f(0) > 0$, the function

$$h_2(z, \lambda) = \lambda f(0) + (1 - \lambda) \tag{2.4.8}$$

gives a homotopy between the constant function $f(0)$ and the identity of A_s. By transitivity, $f(\cdot)$ is homotopic to the identity and hence has a logarithm. $\square$

Now we return to the problem of interpolation by a unit, first studied in Section 2.3. Suppose $\{s_1, \cdots, s_n\}$ is a set of points in C_{+e}, $\{m_1, \cdots, m_n\}$ is a corresponding set of positive integers, and $\{r_{ij}, j = 0, \cdots, m_i - 1, i = 1, \cdots, n\}$ is a corresponding set of complex numbers. The objective is to determine (if possible) a unit u of $\mathbf{S}$ such that

$$u^{(j)}(s_i) = r_{ij}, j = 0, \cdots, m_i - 1, i = 1, \cdots, n . \tag{2.4.9}$$

Using the bilinear transformation $z = (s - 1)/(s + 1)$, one can transform the above problem into an equivalent one of finding a rational unit f of A_s such that

$$f^{(j)}(z_i) = q_{ij}, j = 0, \cdots, m_i - 1, i = 1, \cdots, n , \tag{2.4.10}$$

where $z_i = (s_i - 1)/(s_i + 1)$, $q_{i0} = r_{i0}$, and q_{ij} for $j \geq 1$ is a more complicated function of s_i and of j (see Example 2.4.4). The important point is that z_i is real if and only if s_i is real. One can always renumber the z_i's in such a way that $z_1, \cdots, z_l$ are real and $z_{l+1}, \cdots, z_n$ are nonreal.

Theorem 2.4.3 below shows that there exists a unit f of A_s satisfying (2.4.10) if and only if $q_{10}, \cdots, q_{l0}$ are all of the same sign, and gives a procedure for constructing such a unit when it exists. The unit so constructed is not rational in general, but an indication is also given of a way to construct a rational unit satisfying (2.4.10). Thus Theorem 2.4.3 contains the same result as Theorem 2.3.1, but the proof is considerably simpler.

Theorem 2.4.3   Given elements $z_1, \cdots, z_n$ of **D**, positive integers $m_1, \cdots, m_n$, and complex numbers $q_{ij}, j = 0, \cdots, m_i - 1; i = 1, \cdots, n$, suppose $z_1, \cdots, z_l$ are real and that $z_{l+1}, \cdots, z_n$ are nonreal. Suppose also that q_{ij} is real for all j whenever z_i is real. Under these conditions, there exists a rational unit f of A_s satisfying (2.4.10) if and only if $q_{10}, \cdots, q_{l0}$ are all of the same sign.

Proof. "only if" If f is a unit of A_s, then $f(\sigma)$ does not change sign as σ varies over $[-1, 1]$. Hence if a unit f satisfies (2.4.10) then the constants $q_{10}, \cdots, q_{l0}$ must all have the same sign.

"if" Suppose $q_{10}, \cdots, q_{l0}$ all have the same sign. We may assume without loss of generality that all these constants are positive; if not, the original problem can be replaced by the equivalent problem of finding a unit h that satisfies the interpolation conditions $h^{(j)}(z_i) = -q_{ij} \, \forall i, j$.

It is first shown that there exists a (not necessarily rational) unit $h \in A_s$ satisfying (2.4.10) (with f replaced by h, of course). If we can construct a function $g \in A_s$ satisfying

$$\frac{d^j}{dz^j} \exp(g(z))|_{z=z_i} = q_{ij}, \; \forall i, j \, , \tag{2.4.11}$$

then $h = \exp(g)$ satisfies the conditions (2.4.10). Let us translate (2.4.11) into conditions on the values of g and its derivatives. First,

$$g(z_i) = \log q_{i0}, i = 1, \cdots, n \, . \tag{2.4.12}$$

The noteworthy point is that when $1 \leq i \leq l$ so that z_i is real, q_{i0} is real and positive so that $\log q_{i0}$ is (more precisely, can be chosen to be) real. When z_i is nonreal, $\log q_{i0}$ may be nonreal, but this does not matter. For higher derivatives of g, (2.4.11) leads to

$$g'(z_i) = q_{i1}/q_{i0} \, , \tag{2.4.13}$$
$$g''(z_i) = \{q_{i2} - [g'(z_i)]^2\}/q_{i0} \, , \tag{2.4.14}$$

and so on. Since $q_{i0} \neq 0$ for all i, the expressions in (2.4.13) and (2.4.14) are all well-defined. Moreover, the quantities $g^{(j)}(z_i)$ are all real whenever z_i is real. Thus the original interpolation problem has been reduced to one of constructing a function $g \in A_s$ (*not* required to be a unit) satisfying certain interpolation conditions. It is trivial to construct such a g; in fact, g can always be chosen to be a polynomial using the technique of Lagrange interpolation (see Example 2.4.4). Setting $h = \exp(g)$ gives a unit satisfying (2.4.10).

In general, h need not be rational. To construct a *rational* unit f satisfying (2.4.10), proceed as follows: Let d be any polynomial in A_s such that $d^{(j)}(z_i) = q_{ij}$. Let

$$e(z) = \prod_{i=1}^{n} (z - z_i)^{m_i} . \tag{2.4.15}$$

The fact that $h^{(j)}(z_i) = d^{(j)}(z_i) = q_{ij}$ for all i, j implies that $h - d$ and its first $m_i - 1$ derivatives vanish at z_i, for all i. Hence e divides $h - d$ in A_s. Let $c = (h - d)/e$. Since rational functions (and in fact polynomials) are dense in A_s, there exists a rational function $c_1 \in A_s$ such that $\|c - c_1\| < 1/[\|h^{-1}\| \cdot \|e\|]$. Now define $f = d + c_1 e$. Then f is rational (polynomial if c_1 is polynomial); moreover, $\|h - f\| = \|c - c_1 e\| < 1/\|h^{-1}\|$. Hence f is a unit. Finally, since $d - f = c_1 e$ is a multiple of e and since e has a zero of order m_i at z_i, it follows that

$$f^{(j)}(z_i) = d^{(j)}(z_i) = q_{ij}, \ \forall i, j . \tag{2.4.16}$$

Thus f is rational and satisfies (2.4.10). □

The above proof shows that if $q_{10}, \cdots, q_{l0}$ are all of the same sign, then (2.4.10) can be satisfied by a *polynomial* $f(\cdot)$ with all its zeros outside the unit disc **D** (and all of its poles at infinity). If we define $u(s) = f((s - 1)/(s + 1))$, then it follows that the s-domain conditions (2.4.9) can be satisfied by a unit $u(\cdot) \in$ **S** that has *all* of its poles at $s = -1$ (or any other prespecified point on the negative real axis). Of course, the procedure given in Section 2.3 for constructing u also leads to the same conclusion.

Example 2.4.4 Suppose it is required to construct a unit $u \in$ **S** satisfying

$$u(1) = 1, u'(1) = 0.5, u(j) = 1 + j2, u(\infty) = 2 . \tag{2.4.17}$$

Let $f(z) = u((1 + z)/(1 - z))$. Then f must satisfy

$$f(0) = 1, f'(0) = -1, f(j) = 1 + j2, f(1) = 2 . \tag{2.4.18}$$

Note that $f'(0)$ is computed using the chain rule:

$$f'(0) = \frac{du}{ds}|_{s=1} \cdot \frac{ds}{dz}|_{z=0} . \tag{2.4.19}$$

As in the proof of Theorem 2.4.3, an irrational unit h is constructed that satisfies

$$h(0) = 1, h'(0) = -1, h(j) = 1 + j2, h(1) = 2 . \tag{2.4.20}$$

Let $h = \exp(g)$. In order for h to satisfy (2.4.20), g must satisfy

$$
\begin{aligned}
g(0) &= 0, \; g'(0) = -1 \quad \text{(from (2.4.13))} \\
g(j) &= \log(1 + j2) \approx 0.8 + j1.1 \\
g(1) &= \log 2 \approx 0.7 \; .
\end{aligned}
\tag{2.4.21}
$$

Using Lagrange interpolation, one can readily find a polynomial g satisfying (2.4.21), namely

$$
\begin{aligned}
g(z) &= 0.7z^2(z^2 + 1)/2 + z(z - 1)(z^2 + 1) \\
&\quad + (0.8 + j1.1)z^2(z - 1)(z - j)/(2 + j2) \\
&\quad + (0.8 - j1.1)z^2(z - 1)(z + j)/(2 - 2j) \\
&= 2.3z^4 - 2.1z^3 + 1.5z^2 - z \; .
\end{aligned}
\tag{2.4.22}
$$

Now, $h(z) = \exp(g(z))$ satisfies (2.4.20).

The construction of a *polynomial* unit satisfying (2.4.17) is more complicated.

NOTES AND REFERENCES

The proof that the ring $\mathbf{S}$ is a Euclidean domain is given by Hung and Anderson [50]. Previously Morse [68] had shown that $\mathbf{S}$ is a principal ideal domain. For the related notion of Λ-generalized polynomials, see Pernebo [74]. The interpolation theorem 2.3.1 is given by Youia, Bongiorno and Lu [108], and the current proof is an adaptation of theirs. The more advanced proof of Theorem 2.3.1 given in Section 2.4 is found in Vidyasagar and Davidson [98].

CHAPTER 3

Scalar Systems: An Introduction

The objective of this chapter is to give a glimpse of the power of the factorization approach alluded to in the title of the book, by applying it to the special case of scalar systems (i.e., single-input-single-output systems). Attention is focused on two specific results, namely a parametrization of all compensators that stabilize a given plant, and the development of a necessary and sufficient condition for a plant to be stabilizable by a stable compensator. Proofs are kept to a minimum, especially in cases where they would be routine specializations of more general results for multivariable plants as developed in subsequent chapters. In the same vein, no discussion is given of control problems such as tracking, disturbance rejection, etc., which go beyond mere stabilization; these problems are studied directly in the multivariable case in Chapter 5.

Throughout the chapter, the symbol **S** is used to denote the set of proper stable rational functions, while **U** denotes the set of units of **S**, i.e., functions in **S** whose reciprocal also belongs to **S**.

3.1 PARAMETRIZATION OF ALL STABILIZING COMPENSATORS

Suppose $p \in \mathbb{R}(s)$, so that $p(s)$ is the transfer function of a scalar, lumped, linear, time-invariant system. Suppose a compensator $c \in \mathbb{R}(s)$ is connected to the plant p in the feedback configuration shown in Figure 3.1. The equations describing the closed-loop system are

$$\begin{bmatrix} e_1 \\ e_2 \end{bmatrix} = \begin{bmatrix} u_1 \\ u_2 \end{bmatrix} - \begin{bmatrix} 0 & p \\ -c & 0 \end{bmatrix} \begin{bmatrix} e_1 \\ e_2 \end{bmatrix} , \qquad (3.1.1)$$

which can be solved to yield

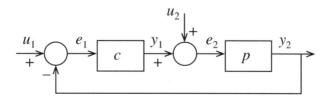

Figure 3.1: Feedback System.

$$\begin{bmatrix} e_1 \\ e_2 \end{bmatrix} = \begin{bmatrix} 1/(1+pc) & -p/(1+pc)) \\ c/(1+pc) & 1/(1+pc)) \end{bmatrix} \begin{bmatrix} u_1 \\ u_2 \end{bmatrix} \tag{3.1.2}$$

provided of course that $1 + pc \neq 0$.[1] Let

$$H(p,c) = \begin{bmatrix} 1/(1+pc) & -p/(1+pc)) \\ c/(1+pc) & 1/(1+pc)) \end{bmatrix} \tag{3.1.3}$$

denote the 2×2 transfer matrix from (u_1, u_2) to (e_1, e_2). Then we say that *the pair (p, c) is stable*, and that *c stabilizes p*, if and only if $H(p, c) \in \mathbf{S}^{2 \times 2}$. Thus, the present notion of stability requires *each* of the four elements of $H(p, c)$ (of which two are the same) to represent a BIBO stable system: it is not enough to have just $p/(1 + pc) \in \mathbf{S}$, which is an oft-stated definition of stability given in undergraduate textbooks. There are two reasons for this: 1) If the compensator c is itself stable, then $H(p, c) \in \mathbf{S}^{2 \times 2}$ if and only if $p/(1 + pc) \in \mathbf{S}$. Hence, if we are using a stable compensator, then closed-loop stability as defined above reduces to the single (and familiar) requirement that $p/(1 + pc)$ be a stable rational function. (See Lemma 5.1.2 for a multivariable version of this fact.) 2) In achieving closed-loop stability, we would like to be sure of having "internal" as well as "external" stability. In other words, it is desirable to insure that, whenever u_1, u_2 are any bounded inputs, *all* resulting signals in the closed-loop system are bounded. It turns out that $H(p, c) \in \mathbf{S}^{2 \times 2}$ is a necessary and sufficient condition to ensure this. (See Lemmas 5.1.1 and 5.1.3 for multivariable versions of this fact.)

Having thus defined closed-loop stability, we next give a necessary and sufficient condition for it.

Lemma 3.1.1   *Suppose $p, c \in \mathbb{R}(s)$, and let $p = n_p/d_p$, $c = n_c/d_c$, where $n_p, d_p, n_c, d_c \in \mathbf{S}$, and n_p, d_p are coprime, n_c, d_c are coprime. Define*

$$\delta(p,c) = n_p n_c + d_p d_c . \tag{3.1.4}$$

Then the pair (p, c) is stable if and only if $\delta(p, c) \in \mathbf{U}$.

Proof. "if" Suppose $\delta(p, c) \in \mathbf{U}$; then $1/\delta(p, c) \in \mathbf{S}$. It is now routine to verify that

$$H(p,c) = \frac{1}{\delta(p,c)} \begin{bmatrix} d_p d_c & -n_p d_c \\ d_p n_c & d_p d_c \end{bmatrix} \tag{3.1.5}$$

belongs to $\mathbf{S}^{2 \times 2}$, so that the pair (p, c) is stable.

"only if" Suppose (p, c) is stable. Then certainly $1 + pc \neq 0$. Also, $d_p \neq 0$, $d_c \neq 0$ since both are denominators of fractions. Hence, $\delta(p, c) = d_p d_c (1 + pc) \neq 0$, and the formula (3.1.5) is valid. Since $H(p, c) \in \mathbf{S}^{2 \times 2}$, it follows from (3.1.5) that

$$\frac{d_p d_c}{\delta(p,c)} \in \mathbf{S}, \ \frac{d_p n_c}{\delta(p,c)} \in \mathbf{S}, \ \frac{n_p d_c}{\delta(p,c)} \in \mathbf{S} . \tag{3.1.6}$$

[1]This means that $1 + pc$ is not the zero function.

Also

$$1 - \frac{d_p d_c}{\delta(p,c)} = \frac{n_p n_c}{\delta(p,c)} \in \mathbf{S} . \tag{3.1.7}$$

A compact way of expressing (3.1.6) and (3.1.7) is

$$\begin{bmatrix} d_p \\ n_p \end{bmatrix} \frac{1}{\delta(p,c)} [d_c \ n_c] \in \mathbf{S}^{2 \times 2} . \tag{3.1.8}$$

Since n_p, d_p are coprime and n_c, d_c are coprime, it follows from Fact A.3.4 that there exist $x_p, y_p, x_c, y_c \in \mathbf{S}$ such that $x_p n_p + y_p d_p = 1, x_c n_c + y_c d_c = 1$. Now (3.1.8) implies that

$$[y_p x_p] \begin{bmatrix} d_p \\ n_p \end{bmatrix} \frac{1}{\delta(p,c)} [d_c \ n_c] \begin{bmatrix} y_c \\ x_c \end{bmatrix} = \frac{1}{\delta(p,c)} \in \mathbf{S} , \tag{3.1.9}$$

which shows that $\delta(p,c) \in \mathbf{U}$. □

Corollary 3.1.2   *Suppose $p \in \mathbb{R}(s)$, and let $p = n_p/d_p$ where $n_p, d_p \in \mathbf{S}$ are coprime. Then $c \in \mathbb{R}(s)$ stabilizes p if and only if $c = n_c/d_c$ for some $n_c, d_c, \in \mathbf{S}$ that satisfy*

$$n_p n_c + d_p d_c = 1 . \tag{3.1.10}$$

Proof. "if" If (3.1.10) holds, then $\delta(p,c) = 1$ which is certainly a unit of $\mathbf{S}$. Hence, (p,c) is stable by Lemma 3.1.1.

"only if" Suppose c stabilizes p, and express c as n_1/d_1 where $n_1, d_1 \in \mathbf{S}$ are coprime. Then, by Lemma 3.1.1, it follows that $\delta = n_1 n_p + d_1 d_p \in \mathbf{U}$. Now define $n_c = n_1/\delta, d_c = d_1/\delta$. Then $n_c, d_c \in \mathbf{S}, c = n_c/d_c$, and n_c, d_c, satisfy (3.1.10). □

The central result of this chapter, which provides a parametrization of *all* compensators that stabilize a given plant, now falls out almost routinely.

Theorem 3.1.3   Suppose $p \in \mathbb{R}(s)$, and let $p = n_p/d_p$ where $n_p, d_p \in \mathbf{S}$ are coprime. Select $x, y \in \mathbf{S}$ such that

$$x n_p + y d_p = 1 . \tag{3.1.11}$$

Then the set of all compensators that stabilize p, denoted by $S(p)$, is given by

$$S(p) = \left\{ c = \frac{x + r d_p}{y - r n_p} : r \in \mathbf{S} \text{ and } y - r n_p \neq 0 \right\} . \tag{3.1.12}$$

Proof. Suppose that c is of the form

$$c = \frac{x + rd_p}{y - rn_p} \tag{3.1.13}$$

for some $r \in \mathbf{S}$. Then, since

$$(x + rd_p) \cdot n_p + (y - rn_p) \cdot d_p = xn_p + yd_p = 1 , \tag{3.1.14}$$

it follows from Corollary 3.1.2 that c stabilizes p.

Conversely, suppose c stabilizes p. Then, from Corollary 3.1.2, $c = n_c/d_c$ where $n_c, d_c \in \mathbf{S}$ satisfy (3.1.10). Thus, the proof is complete if it can be shown that every solution of (3.1.10) must be of the form

$$n_c = x + rd_p, d_c = y - rn_p \tag{3.1.15}$$

for some $r \in \mathbf{S}$. Subtracting (3.1.11) from (3.1.10) and rearranging gives

$$(n_c - x)n_p = (y - d_c)d_p . \tag{3.1.16}$$

Since d_p and n_p are coprime, it follows from Corollary A.3.9 and Problem A.3.11 that d_p divides $(n_c - x)$ and that n_p divides $d_c - y$. Let, r denote $(n_c - x)/d_p$. Then $n_c = x + rd_p$. Now (3.1.16) shows that $d_c = y - rn_p$. □

In order to apply Theorem 3.1.3 to determine the set of *all* compensators that stabilize a given plant p, one needs to do two things: (i) Express p as a ratio n_p/d_p where $n_p, d_p \in \mathbf{S}$ are coprime. (ii) Find a particular solution x, y of (3.1.11). Systematic procedures for achieving these two steps are given in Section 4.2. Note that, in view of Corollary 3.1.2, the second step is equivalent to finding *one* compensator that stabilizes p. For scalar systems, the first step is very easy, as shown in Fact 2.1.3.

Once these two steps are completed, (3.1.12) provides a parametrization of the set $S(p)$ of *all* stabilizing compensators for p. The condition $y - rn_p \neq 0$ is not very restrictive, as $y - rn_p$ can equal zero for at most one choice of r (See Problem 3.1.1).

The utility of Theorem 3.1.3 derives not merely from the fact that it provides a parametrization of $S(p)$ in terms of a "free" parameter r, but also from the simple manner in which this free parameter enters the resulting (stable) closed-loop transfer matrix.

Corollary 3.1.4 *Let the symbols p, n_p, d_p, x, y be as in Theorem 3.1.3. Suppose $c = (x + rd_p)/(y - rn_p)$ where $r \in \mathbf{S}$. Then*

$$H(p, c) = \begin{bmatrix} d_p(y - rn_p) & -n_p(x + rd_p) \\ d_p(x + rd_p) & d_p(y - rn_p) \end{bmatrix} . \tag{3.1.17}$$

Proof. Apply (3.1.5) with $n_c = x + rd_p, d_c = y - rn_p$. □

The application of Theorem 3.1.3 and Corollary 3.1.4 is now illustrated by means of a simple example.

Example 3.1.5   Let

$$p(s) = \frac{s}{(s+1)(s-1)} . \tag{3.1.18}$$

Then $p = n_p/d_p$, where, as in Fact 2.1.3,

$$n_p(s) = \frac{s}{(s+1)^2}, d_p(s) = \frac{(s-1)}{(s+1)} . \tag{3.1.19}$$

Suppose that, using some design procedure, we have found a stabilizing compensator for p, namely

$$c(s) = \frac{2(s+2)}{s-0.5} . \tag{3.1.20}$$

This leads to a particular solution x, y of (3.1.11), as follows: Let

$$n_1(s) = \frac{2(s+2)}{s+1}, d_1(s) = \frac{s-0.5}{s+1} . \tag{3.1.21}$$

Then

$$\delta(p,c) = n_p n_1 + d_p d_1 = \frac{s^2 + 1.5s^2 + 3s + 0.5}{(s+1)^3} . \tag{3.1.22}$$

All zeros of the numerator polynomial of $\delta(p,c)$ are in the open left half-plane, so that $\delta(p,c)$ is a unit of **S**. This is to be expected, since c is a stabilizing compensator. Now, as in the proof of Corollary 3.1.2, define

$$x(s) = \frac{n_1}{\delta(p,c)} = \frac{2(s+2)(s+1)^2}{\phi(s)} , \tag{3.1.23}$$

$$y(s) = \frac{d_1}{\delta(p,c)} = \frac{(s-0.5)(s+1)^2}{\phi(s)} , \tag{3.1.24}$$

where

$$\phi(s) = s^3 + 1.5s^2 + 3s + 0.5 . \tag{3.1.25}$$

Then x, y satisfy (3.1.11). Using Theorem 3.1.3, it now follows that the set of *all* compensators that stabilize p is given, after clearing a few fractions, by those c that are of the form

$$
\begin{aligned}
c(s) &= \frac{x + r d_p}{y - r n_p} \\
&= \frac{2(s+2)(s+1)^4 + r(s)(s-1)(s+1)\phi(s)}{(s-0.5)(s+1)^4 - r(s)s\phi(s)}
\end{aligned}
\tag{3.1.26}
$$

where r is any function in $\mathbf{S}$.[2] The expression (3.1.26) can be simplified considerably. Dividing both numerator and denominator by $(s+1)^3$ gives

$$c(s) = \frac{2(s+2)(s+1) + r(s)\delta(s)(s-1)(s+1)}{(s-0.5)(s+1) - r(s)\delta(s)s}, \qquad (3.1.27)$$

where $\delta(s) = \delta(p,c)(s)$ as defined in (3.1.22). Now $\delta(p,c)$ is a *unit* of $\mathbf{S}$; hence, as r varies freely over $\mathbf{S}$, so does $r\delta$. In other words, the map $r \mapsto r\delta$ is a one-to-one map of $\mathbf{S}$ onto itself. Therefore we may as well replace $r\delta$ by a new "free" parameter q, which is an arbitrary function in $\mathbf{S}$. Thus, the set of compensators that stabilize p is given by

$$S(p) = \left\{ c : c(s) = \frac{2(s+2)(s+1) + q(s)(s-1)(s+1)}{(s-0.5)(s+1) - sq(s)}, q \in \mathbf{S} \right\}. \qquad (3.1.28)$$

From Corollary 3.1.4, the closed-loop transfer matrix $H(p,c)$ corresponding to a compensator c of the form (3.1.27) is given by (3.1.17). To illustrate the manipulations involved, consider the term $h_{11}(p,c)$; it equals

$$d_p(y - rn_p) = \frac{s-1}{s+1} \left\{ \frac{(s-0.5)(s+1)^2}{\phi(s)} - \frac{r(s)s}{(s+1)^2} \right\}$$
$$= \frac{(s-1)[(s-0.5)(s+1) - q(s)s]}{\phi(s)}, \qquad (3.1.29)$$

where $q = r\delta$. After similar manipulations on the other three terms, the final result is

$$h_{21}(p,c)(s) = \frac{(s-1)(s+1)[2(s+2) + q(s)(s-1)]}{\phi(s)},$$
$$h_{12}(p,c)(s) = \frac{s[2(s+2) + q(s)(s-1)]}{\phi(s)}, \qquad (3.1.30)$$

and of course $h_{22}(p,c) = h_{11}(p,c)$. Thus, as q varies over $\mathbf{S}$, the expression (3.1.28) generates *all* compensators that stabilize p, and (3.1.29)–(3.1.30) give the corresponding (stable) closed-loop transfer matrices.

Consider now the problem of stabilizing a given plant p with a general domain of stability $\mathbf{D}$ replacing the open left half-plane. Thus, given $p \in \mathbb{R}(s)$ and a domain of stability $\mathbf{D}$, the problem is to parametrize *all* compensators $c \in \mathbb{R}(s)$ such that the closed-loop transfer matrix $H(p,c)$ of (3.1.3) belongs to $\mathbf{S}_{\mathbf{D}}^{2 \times 2}$. Looking back over the proof of Theorem 3.1.3 and all the preliminary results leading up to it, we see that the entire development depends on just two facts: (i) $\mathbf{S}$ is a proper Euclidean domain, and (ii) every $f \in \mathbb{R}(s)$ can be factorized as a/b where $a, b \in \mathbf{S}$ are coprime. Now, both of these statements remain valid if $\mathbf{S}$ is replaced by $\mathbf{S}_{\mathbf{D}}$. Hence, Theorem 3.1.3 and Corollary 3.1.4 carry over *in toto* if $\mathbf{S}$ is replaced by $\mathbf{S}_{\mathbf{D}}$. This is illustrated by means of an example.

Example 3.1.6   Consider again the plant p of (3.1.18), and let

$$\mathbf{D} = \{s : \text{Re } s < -2\}. \qquad (3.1.31)$$

---

[2]Since $y(\infty) \neq 0$, $n_p(\infty) = 0$ and $r(\infty)$ is finite, we see that $(y - rn_p)(\infty) \neq 0 \, \forall r \in \mathbf{S}$, so that r is unrestricted.

Suppose it is desired to parametrize all compensators c such that the closed-loop transfer matrix $H(p, c)$ is proper and has all of its poles in **D**. The first step is to find $n_p, d_p \in \mathbf{S_D}$ such that $p = n_p/d_p$ and n_p, d_p are coprime. This can be done using the procedure of Fact 2.1.3, except that the term $s + 1$ is replaced by $s + a$ where $a > 2$. Let us take $a = 3$, so that

$$n_p(s) = \frac{s}{(s+3)^2}, d_p(s) = \frac{(s+1)(s-1)}{(s+3)^2}. \qquad (3.1.32)$$

In order to find a particular solution of (3.1.11), the bilinear transformation

$$z = \frac{s-3}{s+3}, s = 3\frac{1+z}{1-z} \qquad (3.1.33)$$

is introduced. This leads to

$$n_p = \frac{1-z^2}{12}, d_p = \frac{2z^2 + 5z + 2}{9}. \qquad (3.1.34)$$

Now, since n_p and d_p are coprime *polynomials* in z, there exist *polynomials* x, y in z such that $xn_p + yd_p = 1$. One such pair x, y can be found by first performing the Euclidean division algorithm on n_p and d_p and then back-substituting. This gives

$$x = \frac{10z + 17}{9} = \frac{9s - 7}{3(s+3)}, \qquad (3.1.35)$$

$$y = 5z - 4 = \frac{s - 27}{s+3}. \qquad (3.1.36)$$

The corresponding nominal (or starting) stabilizing compensator[3] is

$$c = \frac{x}{y} = \frac{9s - 7}{3(s - 27)}. \qquad (3.1.37)$$

The set of *all* compensators that stabilize p, in the sense that $H(p, c) \in \mathbf{S_D^{2\times2}}$, is given by

$$S_\mathbf{D}(p) = \left\{ c : c = \frac{x + rd_p}{y - rn_p} \text{ for some } r \in \mathbf{S_D} \right\}, \qquad (3.1.38)$$

and the corresponding $H(p, c)$ is given by (3.1.17).

Now suppose the performance specifications on the closed-loop system are made more stringent, as follows: Define

$$\mathbf{D}_1 = \{s : \text{ Re } s < -2, |\text{ Im } s| \leq |\text{ Re } s|\}. \qquad (3.1.39)$$

The region $\mathbf{D}_1$ is shown in Figure 3.2.

[3]Recall that the phrase "stabilize" has here the meaning "place all closed-loop poles in the prescribed domain of stability."

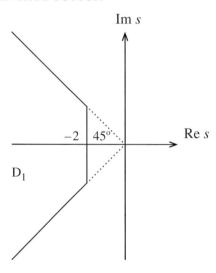

Figure 3.2: Generalized Region of Stability.

Suppose it is desired to parametrize all compensators c such that $H(p, c)$ is proper and has all of its poles inside the region $\mathbf{D}_1$ (rather than just $\mathbf{D}$ as in the previous paragraph). To obtain such a parametrization, it is necessary first to obtain a coprime pair $n_p, d_p \in \mathbf{S}_{\mathbf{D}_1}$ and then to find a pair $x, y \in \mathbf{S}_{\mathbf{D}_1}$ such that $xn_p + yd_p = 1$. Fortunately, the functions n_p, d_p, x, y in (3.1.32), (3.1.35), (3.1.36) already satisfy this requirement. Hence, the desired parametrization is

$$\mathbf{S}_{\mathbf{D}_1}(p) = \left\{ c : c = \frac{x + rd_p}{y - rn_p} \text{ for some } r \in \mathbf{S}_{\mathbf{D}_1} \right\}. \tag{3.1.40}$$

Note that the expressions in (3.1.38) **and** (3.1.39) **are the same**; the only difference is that the "free" parameter r in (3.1.39) is only permitted to range over the smaller set $\mathbf{S}_{\mathbf{D}_1}$, whereas in (3.1.38) it is permitted to range over the larger set $\mathbf{S}_{\mathbf{D}}$.

PROBLEMS

3.1.1. Given that $a, b, c, d \in \mathbf{S}$ satisfy $ab + cd = 1$, show that $a - rc$ can equal zero for at most one $r \in \mathbf{S}$. (Hint: If $c \neq 0$, then $a - rc = 0$ implies that $a - qc \neq 0$ for all $q \neq r$. If $c = 0$ then $a \neq 0$.) Construct $a, b, c, d \in \mathbf{S}$ such that $ab + cd = 1$ and c divides a (which is equivalent to saying that $a - rc = 0$ for some r). Describe *all* such c.

3.1.2. Suppose $a, b, c, d \in \mathbf{S}$ satisfy $ab + cd = 1$. Show that if c is not a unit then c does not divide a (i.e., $a - rc \neq 0 \, \forall r$). Apply this result to Theorem 3.1.3 to show that if the plant p is strictly proper, then $y - rn_p \neq 0 \, \forall r \in \mathbf{S}$.

3.1.3. In Theorem 3.1.3, show that distinct choices of the parameter r lead to distinct compensators.

3.1.4. Find all compensators that stabilize the plant

$$p(s) = \frac{s-1}{(s-2)(s^2+1)} \, .$$

3.1.5. Given the BIBO stable plant

$$p(s) = \frac{s^2+1}{(s+1)^2} \, ,$$

find all compensators c such that $H(p, c) \in \mathbf{S}_{\mathbf{D}_1}^{2 \times 2}$, where $\mathbf{D}_1$ is as in (3.1.39).

3.2   STABILIZATION USING A STABLE COMPENSATOR

A plant $p \in \mathbb{R}(s)$ is said to be *strongly stabilizable* if there exists a $c \in \mathbf{S}$ such that (p, c) is stable, i.e., if p can be stabilized using a stable compensator. A study of strong stabilizability is important for at least two reasons: (i) Suppose p is strongly stabilizable. Let $c \in \mathbf{S}$ stabilize p and let $p_1 = p/(1 + pc)$ denote the resulting plant transfer function. Since c is stable, a factorization of c is given by $n_c = c, d_c = 1$. It now follows from (3.1.5) that $p_1 = n_p$ where n_p is the numerator of a factorization of p. In other words, if p is stabilized using a *stable* compensator, then the resulting plant transfer function $p_1 = p/(1 + pc)$ has the same C_{+e}-zeros as p *and no others*. The argument is reversible: If p is *not* strongly stabilizable, then d_c in (3.1.5) can never be a unit, so that p_1 always has some additional C_{+e}-zeros beyond those of p. Since the C_{+e}-zeros of p_1 affect its sensitivity to disturbances and its ability to track reference inputs (see Section 5.6), it is desirable not to introduce, in the process of stabilization, additional C_{+e}-zeros into the closed-loop transfer function beyond those of p. But this is possible if and only if p is strongly stabilizable. (ii) Several problems in compensator design, such as simultaneous stabilization (Section 5.4) and two-step stabilization (Section 5.2) are closely related to strong stabilizability. Accordingly, in this section necessary and sufficient conditions are derived for a plant to be strongly stabilizable. These conditions turn out to be very easily stated in terms of the locations of the real C_{+e}-zeros and poles of the plant under study.

Theorem 3.2.1   Given $p \in \mathbb{R}(s)$, let $\sigma_1, \cdots, \sigma_l$ denote the real C_{+e}-zeros of p (including ∞ if p is strictly proper), arranged in ascending order. Let η_i denote the number of poles of p (counting multiplicity) in the interval (σ_i, σ_{i+1}), and let η denote the number of odd integers in the sequence $\eta_1, \cdots, \eta_{l-1}$. Then every c that stabilizes p has at least η poles in C_{+e}. Moreover, this lower bound is exact in that there is a c that stabilizes p that has exactly η C_{+e}-poles.

Corollary 3.2.2   *A plant $p \in \mathbb{R}(s)$ is strongly stabilizable if and only if the number of poles of p between every pair of real C_{+e}-zeros of p is even.*

The property described in Corollary 3.2.2 is called the *parity interlacing property (p.i.p.)*. Thus, a plant p is strongly stabilizable if and only if it has the p.i.p.

Proof of Theorem 3.2.1. By Theorem 3.1.3, every c that stabilizes p has the form $(x + rd_p)/(y - rn_p)$ for some $r \in \mathbf{S}$, where all symbols are defined in Theorem 3.1.3. Further, $x + rd_p$ and $y - rn_p$, are coprime, by (3.1.14). Hence, by Fact 2.1.3, the number of C_{+e}-poles of c equals $\delta(y - rn_p)$, i.e., the degree of $y - rn_p$ in the Euclidean domain $\mathbf{S}$. Now a lower bound on this degree as r varies can be computed using Theorem 2.3.2. Again by Fact 2.1.3, the C_{+e}-zeros of n_p are precisely $\sigma_1, \cdots, \sigma_l$. Thus, by Theorem 2.3.2, $\min_{r \in \mathbf{S}} \delta(y - rn_p)$ equals the number of sign changes in the sequence $\{y(\sigma_1), \cdots, y(\sigma_l)\}$. Next, observe that, since $xn_p + yd_p = 1$, it follows that $y(\sigma_i)d_p(\sigma_i) = 1\ \forall i$ (since $n_p(\sigma_i) = 0$). Hence, $y(\sigma_i)$ and $d_p(\sigma_i)$ have the same sign for all i. Thus, $\min_{r \in \mathbf{S}} \delta(y - rn_p)$ also equals the number of sign changes in the sequence $\{d_p(\sigma_1), \cdots, d_p(\sigma_l)\}$. To conclude the proof, note that $d_p(\sigma_{i+1})$ has the same (resp. opposite) sign as $d_p(\sigma_i)$ if and only if the number of zeros of d_p in the interval (σ_i, σ_{i+1}) is even (resp. odd). Since the C_{+e}-zeros of d_p are same as the poles of p (Fact 2.1.3), the result follows. $\qquad\square$

The proof of the corollary is omitted as it is obvious.

The following alternate form of Corollary 3.2.2 is sometimes useful.

Corollary 3.2.3 *Suppose $p \in \mathbb{R}(s)$ equals n_p/d_p where $n_p, d_p \in \mathbf{S}$ are coprime. Then p is strongly stabilizable if and only if $d_p(\cdot)$ has the same sign at all real C_{+e}-zeros of p (or equivalently, n_p).*

Once the strong stabilizability of a given plant p is established using Corollary 3.2.2 or Corollary 3.2.3, the construction of a stable stabilizing compensator for p is straight-forward using the results of Sections 2.3 or 2.4. Specifically, factor p as n_p/d_p where $n_p, d_p \in \mathbf{S}$ are coprime. Then the conditions of either corollary assure the existence of a unit u that interpolates the values of d_p, at the C_{+e}-zeros of n_p; moreover, such a unit can be constructed using the procedure described in the proof of Theorem 2.3.1 or Theorem 2.4.3. Once such a u is constructed, the stable stabilizing compensator is given by $c = (u - d_p)/n_p$.

Example 3.2.4 Consider again the plant

$$p(s) = \frac{s}{(s+1)(s-1)}. \tag{3.2.1}$$

This plant has two real C_{+e}-zeros, namely $\sigma_1 = 0, \sigma_2 = \infty$. Since p has one pole in the interval $(0, \infty)$, it is *not* strongly stabilizable.

Example 3.2.5 Consider the plant

$$p(s) = \frac{s(s-0.4)}{(s+1)(s-1)(s-2)}. \tag{3.2.2}$$

By applying the criterion of Corollary 3.2.2, we see that p is strongly stabilizable. To construct a stable c that stabilizes p, we proceed as follows: First, factorize p as a ratio n_p/d_p where $n_p, d_p \in \mathbf{S}$ are coprime. This is most easily done using Fact 2.1.3, which leads to

$$n_p(s) = \frac{s(s - 0.4)}{(s + 1)^3}, d_p(s) = \frac{(s - 1)(s - 2)}{(s + 1)^2} . \tag{3.2.3}$$

The problem of constructing a stable stabilizing compensator is equivalent to constructing a unit $u \in \mathbf{S}$ that interpolates the values of d_p at the C_{+e}-zeros of n_p. Thus, the problem is one of constructing a unit $u \in \mathbf{S}$ that satisfies

$$u(0) = 2, u(0.4) = \frac{24}{49}, u(\infty) = 1 . \tag{3.2.4}$$

Let us find such a unit using the iterative procedure contained in the proof of Theorem 2.3.1. First, let $u_0(s) \equiv 1$; this unit interpolates the required value at infinity. Next, to meet the constraint at 0.4, define $f_1(s) = 1/(s + 1)$, and let

$$u_1 = (1 + bf_1)^r u_0 , \tag{3.2.5}$$

where the constant b and the integer r are to be selected such that $|b| < \|f_1\|^{-1}$ and $u_1(0.4) = 24/49$. Routine calculations show that $\|f_1\| = 1$, so that $b = 5/7, r = 1$ are suitable, leading to

$$u_1(s) = \frac{7s + 2}{7s + 7} . \tag{3.2.6}$$

This unit meets the interpolation constraints at $s = 0.4$ and $s = \infty$. To meet the last interpolation constraint, define

$$f_2(s) = \frac{s - 0.4}{(s + 1)^2} ,$$
$$u_2 = (1 + bf_2)^r u_1 , \tag{3.2.7}$$

where b must satisfy $|b| < \|f_2\|^{-1}$ and r is an integer. In the present instance this translates to $|b| < 2.0$. Again it is routine to verify that $r = 4$ is a suitable choice, leading to $b = (1 - 7^{1/4})/0.4 \approx 1.5165$, and

$$1 + bf_2 = \frac{s^2 + 0.4835s + 1.6266}{s^2 + 2s + 1} ,$$
$$u_2 = u(s) = \left\{ \frac{s^2 + 0.4835s + 1.6266}{s^2 + 2s + 1} \right\}^4 \frac{7s + 2}{7s + 7} . \tag{3.2.8}$$

Finally, the compensator is given by $c = (u - d_p)/n_p$.

Next, consider the problem of strong stabilizability with a general domain of stability $\mathbf{D}$. Thus, the problem is whether a given plant $p \in \mathbb{R}(s)$ can be "stabilized" by a compensator $c \in \mathbf{S_D}$ such that $H(p, c) \in \mathbf{S_D^{2 \times 2}}$. If the complement of the region $\mathbf{D}$ is simply connected, then both Theorem 3.2.1

and Corollary 3.2.2 apply, the only modification being that C_{+e} is replaced by "the union of $\{\infty\}$ and the complement of $\mathbf{D}$."

Example 3.2.6   Suppose

$$p(s) = \frac{(s+1)(s-1)(s^2+1)}{s(s-2)^2[(s-1)^2+1]} . \tag{3.2.9}$$

Then p has two *real* C_{+e}-zeros, namely $\sigma_1 = 1$ and $\sigma_2 = \infty$. Since p has two poles in the interval $(1, \infty)$, namely the double pole at $s = 2$, p is strongly stabilizable.

Now let

$$\mathbf{D} = \{s : \text{ Re } s < -2| \text{ Im } s| \leq 2| \text{ Re } s|\} . \tag{3.2.10}$$

Then p has three real zeros in the extended complement of $\mathbf{D}$, namely $\sigma_1 = -1, \sigma_2 = 1, \sigma_3 = \infty$. In the interval $(-1, 1)p$ has one pole. Hence, no $c \in \mathbf{S_D}$ exists such that $H(p, c) \in \mathbf{S_D^{2\times2}}$, i.e., p is *not* strongly stabilizable with respect to the domain of stability $\mathbf{D}$.

Finally, consider the following generalization of the strong stabilization problem: Given a plant p and a general domain of stability $\mathbf{D}$, find a compensator $c \in \mathbf{S}$ such that the closed-loop transfer matrix $H(p, c) \in \mathbf{S_D^{2\times2}}$. Thus, the objective is to design a stable compensator such that the closed-loop performance meets more stringent requirements than just stability with respect to the left half-plane. Note that the compensator is merely required to be stable in the conventional sense. The solution to this problem is given without proof, and is this: Such a compensator exists if and only if the plant p satisfies the parity interlacing property with respect to the region C_{+e}, i.e., p is strongly stabilizable in the conventional sense. Applying this test to the plant p of Example 3.2.6, we see that all closed-loop poles can be placed in the region $\mathbf{D}$ using a stable compensator, even though p is not strongly stabilizable with respect to the region $\mathbf{D}$.

The case of discrete-time systems can be handled in a straight-forward manner. Indeed, the contents of Section 2.4 make clear that the strong stabilization problem is most naturally solved by employing a bilinear transformation to map the extended right half-plane into the closed unit disc. Thus, a discrete-time system p is strongly stabilizable if and only if it satisfies the parity interlacing property with respect to the interval $[-1, 1]$.

PROBLEMS

3.2.1. Determine whether each of the strictly proper plants in Figure 3.3 satisfies the parity interlacing property.

3.2.2. Find a stable stabilizing compensator for the plant

$$p(s) = \frac{s(s-1)}{(s^2+1)(s-2)(s-3)} .$$

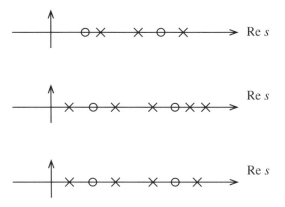

Figure 3.3:

3.3   PARAMETRIZATION OF ALL STABLE STABILIZING COMPENSATORS

Given a plant $p \in \mathbb{R}(s)$, the results of the previous section can be used to determine whether or not p can be stabilized using a stable compensator, i.e., whether or not p is strongly stabilizable. If p is strongly stabilizable, the methods of Sections 2.3 and 2.4 can be used to construct *a* stable stabilizing compensator for p. The objective of this section is to parametrize *all* such compensators.

After the bilinear transformation introduced in Section 2.4, the problem under study can be formulated as follows: Given rational functions d and n in the Banach algebra A_s, find all rational $c \in A_s$ such that $d + cn$ is a unit of A_s. Clearly this is equivalent to finding *all* units $u \in A_s$ such that $d - u$ is a multiple of n. Let $z_1, \cdots, z_l$ be the zeros of n in the unit disc **D**, with multiplicities $m_1, \cdots, m_l$. Then the problem is one of finding all units u in A_s that satisfy

$$u^{(j)}(z_i) = d^{(j)}(z_i), 0 \le j \le m_i - 1; 1 \le i \le l . \tag{3.3.1}$$

Suppose u_1 and u_2 are two units that each satisfy (3.3.1), and let v denote the unit $u_2 u_1^{-1}$. Then a routine calculation shows that

$$v(z_i) = 1, 1 \le i \le l ; \tag{3.3.2}$$
$$v^{(j)}(z_i) = 0, 1 \le j \le m_i - 1; 1 \le i \le l . \tag{3.3.3}$$

Equivalently, v is of the form $1 + fn$ for some $f \in A_s$, since v interpolates the constant function "1" and its derivatives at the zeros of n. This leads to the next result.

Proposition 3.3.1   *Given $d, n \in A_s$, let u be any unit in A_s satisfying (3.3.1). Then the set of all units satisfying (3.3.1) is given by $\{uv : v \in A_s$ is a unit of the form $1 + fn$ for some $f \in A_s\}$.*

For a given $g \in A_s$, let $\mathbf{U}(g)$ denote the set of all units of the form $1 + fg$ where $f \in A_s$. If we can parametrize $\mathbf{U}(g)$, then Proposition 3.3.1 enables us to find all units that satisfy (3.3.1). Then the set of all $c \in A_s$, that stabilize the plant n/d is just $\{(u - d)/c : u$ satisfies $(3.3.1)\}$.

The next two results give an explicit description of $\mathbf{U}(g)$ given a $g \in A_s$. It is necessary to treat separately the cases where g has real zeros in $\mathbf{D}$ and where it does not. To aid in the presentation of the results, some notation is introduced. Suppose $g \in A_s$ has only a finite number of zeros in $\mathbf{D}$. Let $z_1, \cdots, z_s, \bar{z}_1, \cdots, \bar{z}_s, \sigma_1, \cdots, \sigma_r$ denote the distinct zeros of g in $\mathbf{D}$, where $z_1, \cdots, z_s$ are nonreal and $\sigma_1, \cdots, \sigma_r$ are real. For convenience, let $z_1, \cdots, z_{2s+r}$ denote the same sequence, and let μ_i denote the multiplicity of z_i as a zero of g. Select polynomials $p_1(z), \cdots, p_{2s}(z)$ such that

$$p_i(z_i) = 1,\ p_i^{(j)}(z_i) = 0 \text{ for } j = 1, \cdots, \mu_i - 1 \ ,$$
$$p_i^{(j)}(z_k) = 0 \text{ for } j = 0, \cdots, \mu_i - 1 \text{ if } k \neq i \ . \tag{3.3.4}$$

Since the zeros of g occur in complex conjugate pairs, we may suppose that $p_i(z) = \bar{p}_{i+s}(\bar{z})$ for $i = 1, \cdots, s$. Finally, define the polynomials

$$\phi_i(z) = j[p_i(z) - p_{i+s}(z)], i = 1, \cdots, s \ , \tag{3.3.5}$$

and observe that $\phi_i \in A_s\ \forall i$.

Proposition 3.3.2   *Suppose $g \in A_s$ has only a finite number of zeros in $\mathbf{D}$, and that g has at least one real zero in $\mathbf{D}$. Then every unit in A_s of the form $1 + fg$, $f \in A_s$, can be expressed as $\exp(v)$, where $v \in A_s$ has the form*

$$v = hg + \sum_{i=1}^{s} 2\pi m_i \phi_i \ , \tag{3.3.6}$$

where $h \in A_s, m_1, \cdots, m_s$ are arbitrary integers, and ϕ_i is defined in (3.3.5). Conversely, every unit $\exp(v)$ where v is of the form (3.3.6) can also be written as $1 + fg$ for some $f \in A_s$. In summary

$$\mathbf{U}(g) = \{\exp(v) : v \text{ is of the form (3.3.6)} \} \ . \tag{3.3.7}$$

Proof. Suppose $u = 1 + fg$ is a unit. Since g has at least one real zero in $\mathbf{D}$, $u(z)$ equals one for some real z in $\mathbf{D}$. Hence, $u(z) > 0$ for $z \in [-1, 1]$, and by Lemma 2.4.2, u has a logarithm v in A_s. Since $g(z_i) = 0\ \forall i$, it follows that

$$v(z_i) = j2\pi m_i\ \forall i \ , \tag{3.3.8}$$

where m_i is some integer. Since $v(z) = \bar{v}(\bar{z})\ \forall z$, it is immediate from (3.3.8) that $m_i = -m_{i+s}$, for $i = 1, \cdots, s$, and that $m_i = 0$ for $i = 2s + 1, \cdots, 2s + r$. At multiple zeros of g, successive higher derivatives of g vanish, which implies that the corresponding derivatives of u and v also vanish. Thus,

$$v^{(j)}(z_i) = 0 \text{ for } j = 1, \cdots, \mu_i - 1; i = 1, \cdots, 2s + r \ . \tag{3.3.9}$$

Now (3.3.8) and (3.3.9) lead to the conclusion that g divides the function $v - \sum_{i=1}^{2s} j2\pi m_i p_i$, which equals $v - \sum_{i=1}^{s} 2\pi m_i \phi_i$, This is precisely (3.3.6).

Conversely, suppose v is of the form (3.3.6). Then v satisfies (3.3.8) and (3.3.9), which in turn implies that $\exp(v) - 1$ is divisible by g. Hence, $\exp(v)$ is of the form $1 + fg$ for some $f \in A_s$. □

In practice, the assumption that g has at least one real zero in $\mathbf{D}$ does not pose a major restriction, because if g is the "numerator" of a strictly proper plant, then $g(1) = 0$. Also, note that g is not required to be rational - merely to have only a finite number of zeros in $\mathbf{D}$.

We state the next result in the interests of completeness. The proof is omitted, as it closely follows that of Proposition 3.3.2.

Proposition 3.3.3   *Suppose $g \in A_s$ has only a finite number of zeros in $\mathbf{D}$, all of them nonreal. Then every unit $u \in A_s$ of the form $1 + fg$, $f \in A_s$, can either be written as $\exp(v)$ where*

$$v = hg + \sum_{i=1}^{s} 2\pi m_i \phi_i ,\tag{3.3.10}$$

or else as $-\exp(v)$ where

$$v = hg + \sum_{i=1}^{s} (2m_i + 1)\pi \phi_i ,\tag{3.3.11}$$

where $h \in A_s, m_1, \cdots, m_s$, are arbitrary integers, and ϕ_i is defined in (3.3.5). Conversely, every unit $\exp(v)$ where v is of the form (3.3.10) and every unit $-\exp(v)$ where v is of the form (3.3.11) can be written as $1 + fg$ for some $f \in A_s$.

Example 3.3.4   Consider the problem of determining all stable stabilizing compensators for the plant whose transfer function is

$$p(s) = \frac{s^2 + 1}{(s + 2)^2(s - 3)} .$$

After substituting $s = (1 + z)/(1 - z)$, this becomes

$$\frac{n(z)}{d(z)} = \frac{(z^2 + 1)(1 - z)}{(z - 3)^2(2z - 1)}$$

where n and d denote the numerator and denominator polynomials, respectively. Thus, the problem is one of finding all $c \in A_s$, such that $d + cn$ is a unit of A_s. This can be solved using Propositions 3.3.1

and 3.3.2. First, we construct a unit u_0 such that $u_0 - d$ is a multiple of n. Since the zeros of n inside the unit disc are at $1, \pm j$, we require $u_0(z)$ to equal $d(z)$ at these points. Thus, we must have

$$u_0(1) = 4, u_0(j) = 4 + j22 .$$

If $u_0 = \exp(f_0)$, then f_0 must satisfy

$$f_0(1) = \log 4 \approx 1.4, f_0(j) = \log(4 + j22) \approx 3.1 + j1.4 .$$

Using Lagrange interpolation, one can find

$$f_0(z) = -1.55z^2 + 1.4z + 1.55 .$$

Therefore, *one* stabilizing compensator is given by

$$c_0(z) = [d(z) - \exp(f_0(z))]/n(z) .$$

To find *all* stabilizing compensators, we compute the set $\mathbf{U}(n)$. Again using Lagrange interpolation, one can find polynomials to satisfy (3.3.4), namely

$$p_1(z) = \frac{(z-1)(z+j)}{2 + j2}, \; p_2(z) = \bar{p}_1(\bar{z}) ,$$
$$\phi_1(z) = j[p_1(z) - p_2(z)] = -\frac{\pi}{2}(z-1)^2 .$$

Thus, $\mathbf{U}(n) = \{\exp(v)\}$ where $v = hn + \pi m(z-1)^2$, m an arbitrary integer and $h \in A_s$. Therefore the set of all stable stabilizing compensators is given by

$$\left\{ \exp[(f(z)) - d(z)]/n(z) \text{ where} \right.$$
$$\left. f(z) = f_0(z) + h(z)n(z) + \pi m(z-1)^2, m \in \mathbf{Z}, h \in A_s \right\} .$$

NOTES AND REFERENCES

The parametrization of all compensators that stabilize a given plant is given in [109], and in the present form, in [25]. The necessity and sufficiency of the parity interlacing property for strong stabilizability is shown in [108], and its extension as stated in Theorem 3.2.1 is contained in [102]. Anderson [2] gives a method for testing the p.i.p. without computing the poles and zeros of a plant. The results of Section 3.3 are from [98].

CHAPTER 4

Matrix Rings

In the preceding chapter, we have seen the solution of some important problems in controller design for *scalar* systems. The objective of this chapter is to develop the machinery that will enable us to tackle *multivariable* systems. Accordingly, several problems are formulated and solved in the set of *matrices* over a commutative ring. These problems are posed in a fairly general setting so as to unify the controller synthesis techniques for continuous-time as well as discrete-time systems within a single framework.

Throughout this chapter, $\mathbf{R}$ denotes a commutative domain with identity and $\mathbf{F}$ denotes the field of fractions associated with $\mathbf{R}$. Through most of the chapter, $\mathbf{R}$ is merely assumed to be a principal ideal domain. At various places, $\mathbf{R}$ is further restricted to be a proper Euclidean domain.

Throughout this chapter we shall be dealing with matrices whose elements belong to either $\mathbf{R}$ or $\mathbf{F}$. Much of the time, the actual orders of the matrices under study are unimportant to the concepts being put forward. For this reason, $\mathbf{M}(\mathbf{R})$ is used as a generic symbol to denote the set of matrices with elements in $\mathbf{R}$, of whatever order. Thus, "$A, B \in \mathbf{M}(\mathbf{R})$" means that A, B are matrices whose elements are in $\mathbf{R}$, but does *not* mean that A, B have the same order. In cases where it is necessary to display explicitly the order of a matrix, a notation of the form "$A \in \mathbf{R}^{n \times m}$" is used to indicate that A is an $n \times m$ matrix with elements in $\mathbf{R}$. The symbols $\mathbf{M}(\mathbf{F})$ and $\mathbf{F}^{n \times m}$ are defined analogously. Finally, $\mathbf{U}$ denotes the set of units in $\mathbf{R}$, and $\mathbf{U}(\mathbf{R})$ denotes the set of unimodular matrices (of whatever order) in $\mathbf{M}(\mathbf{R})$. The symbol $\mathbf{U}(\mathbf{F})$ is never used.

Throughout, I denotes the identity matrix. If appropriate, I_m is used to denote the identity matrix of order $m \times m$.

4.1 COPRIME FACTORIZATIONS OVER A PRINCIPAL IDEAL DOMAIN

Suppose $\mathbf{R}$ is a principal ideal domain (p.i.d.), $\mathbf{F}$ is the field of fractions associated with $\mathbf{R}$, and that a/b is a fraction in $\mathbf{F}$. Then it is always possible to express a/b as an equivalent fraction f/g where f, g are coprime (i.e., their greatest common divisor is 1; see Fact A.3.6). The main objective of this section is to develop analogous results for matrices in $\mathbf{M}(\mathbf{F})$.

We begin by defining the concepts of a multiple, common divisor, and coprime pair for the set $\mathbf{M}(\mathbf{R})$. Since matrix multiplication is in general noncommutative, it is necessary to make a distinction between *left* multiple and *right* multiple, etc.

Definition 4.1.1 Suppose $A \in \mathbf{M}(\mathbf{R})$; then a *square* matrix $D \in \mathbf{M}(\mathbf{R})$ is a *right divisor* of A, and A is a *left multiple* of D, if there is a $C \in \mathbf{M}(\mathbf{R})$ such that $A = CD$. Suppose $A, B \in \mathbf{M}(\mathbf{R})$ and have the same number of columns. Then a *square matrix* $D \in \mathbf{M(R)}$ is a *greatest common right divisor (g.c.r.d.)* of A, B if

(GCRD1) D is a right divisor of both A and B, and
(GCRD2) D is a left multiple of every common right divisor of A, B.

Two matrices $A, B \in \mathbf{M(R)}$ having the same number of columns are *right-coprime* if every g.c.r.d. of A and B is unimodular.

The definitions of left divisor, g.c.l.d., and left-coprimeness are entirely analogous.

The next result is a matrix version of Fact A.3.2.

Lemma 4.1.2   *Suppose, $A, B \in \mathbf{M(R)}$ and D is a g.c.r.d. of A, B; then so is UD whenever $U \in \mathbf{U(R)}$. If D_1 is another g.c.r.d. of A, B, then there exists a $U \in \mathbf{U(R)}$ such that $D_1 = UD$. Thus, if D is any g.c.r.d. of A, B, then the set of all g.c.r.d.'s of A, B is given by $\{UD : U \in \mathbf{U(R)}\}$.*

Proof. [1] The first sentence is easily proved. To prove the second sentence, observe that since D and D_1 are both g.c.r.d.'s of A, B, there exist square matrices $V, W \in \mathbf{M(R)}$ such that $D_1 = VD, D = WD_1$. If $|D| \neq 0$, the rest of the proof is easy. Since $D = WVD$, it follows that $|V| \cdot |W| = 1$, whence both V and W are unimodular. To complete the proof, we must address the case where $|D| = 0$.

If $|D| = 0$, then the relation $D = WVD$ does not determine V and W uniquely. Thus, the claim is that for some unimodular U, we have $D_1 = UD$ (though D_1 could also equal VD for some nonunimodular matrix V; see Problem 4.1.1). The first step in the proof is to observe that since D_1 is a left multiple of D, we have rank $(D_1) \leq \text{rank}(D)$. By symmetry, rank $(D_1) \leq \text{rank}(D)$, which implies that D_1 and D have the same rank. To be specific, suppose D, D_1 have rank r and are of order $n \times n$. By Theorem B.2.2 there exist $Y, Y_1 \in \mathbf{R}^{n \times n}$, $H, H_1, \in \mathbf{R}^{r \times n}$ such that

$$YD = \begin{bmatrix} H \\ 0 \end{bmatrix} =: \bar{H}, \quad Y_1 D_1 = \begin{bmatrix} H_1 \\ 0 \end{bmatrix} =: \bar{H}_1 , \tag{4.1.1}$$

where Y, Y_1 are unimodular and H, H_1 have rank r. Since D is a left multiple of D_1 and vice versa, it follows from (4.1.1) that $\bar{H}$ is a left multiple of $\bar{H}_1$ and vice versa. Suppose $\bar{H} = R\bar{H}_1, \bar{H}_1 = S\bar{H}$, and partition R, S appropriately. Then $\bar{H} = R\bar{H}_1$ implies that

$$\begin{bmatrix} H \\ 0 \end{bmatrix} = \begin{bmatrix} R_{11} & R_{12} \\ R_{21} & R_{22} \end{bmatrix} \begin{bmatrix} H_1 \\ 0 \end{bmatrix} . \tag{4.1.2}$$

Now, since H_1 has full rank, $R_{21} H_1 = 0$ implies that $R_{21} = 0$. Also, R_{12} and R_{22} are multiplied by the zero matrix and can therefore be replaced by any arbitrary matrices without affecting the validity of (4.1.2). In particular,

$$\begin{bmatrix} H \\ 0 \end{bmatrix} = \begin{bmatrix} R_{11} & 0 \\ 0 & I_{n-r} \end{bmatrix} \begin{bmatrix} H_1 \\ 0 \end{bmatrix} =: \bar{R}\bar{H}_1 . \tag{4.1.3}$$

---

[1] Proof due to Chris Ma.

Similarly,

$$\begin{bmatrix} H_1 \\ 0 \end{bmatrix} = \begin{bmatrix} S_{11} & 0 \\ 0 & I_{n-r} \end{bmatrix} \begin{bmatrix} H \\ 0 \end{bmatrix} =: \bar{S}\bar{H} . \tag{4.1.4}$$

Now (4.1.3) and (4.1.4) together imply that $\bar{H} = \bar{R}\bar{S}\bar{H}$, and in particular, that $H = R_{11} S_{11} H$. Since H has full row rank, it follows that $R_{11} S_{11} = I_r$. Thus, $R_{11}, S_{11} \in \mathbf{U}(\mathbf{R})$, whence $\bar{R}, \bar{S} \in \mathbf{U}(\mathbf{R})$. Finally, it follows from (4.1.1) that $D_1 = Y_1^{-1}\bar{H}_1 = Y_1^{-1}\bar{S}Y^{-1}D = UD$, where $U := Y_1^{-1}\bar{S}Y^{-1} \in \mathbf{U}(\mathbf{R})$. $\qquad\square$

Theorem 4.1.3   Suppose $A, B \in \mathbf{M}(\mathbf{R})$ have the same number of columns. Then A, B have a g.c.r.d. Moreover, if D is any g.c.r.d. of A, B, then there exist $X, Y \in \mathbf{M}(\mathbf{R})$ such that

$$XA + YB = D . \tag{4.1.5}$$

Proof. Define $F \in \mathbf{M}(\mathbf{R})$ by

$$F = \begin{bmatrix} A \\ B \end{bmatrix} , \tag{4.1.6}$$

and consider first the case where F has at least as many rows as columns. By the existence of the Hermite form (Corollary B.2.5), there is a matrix $T \in \mathbf{U}(\mathbf{R})$ such that

$$TF = \begin{bmatrix} D \\ 0 \end{bmatrix} , \tag{4.1.7}$$

where D is upper triangular (though this is not needed in the proof). Partitioning T appropriately and recalling the definition of F gives

$$\begin{bmatrix} T_{11} & T_{12} \\ T_{21} & T_{22} \end{bmatrix} \begin{bmatrix} A \\ B \end{bmatrix} = \begin{bmatrix} D \\ 0 \end{bmatrix} , \tag{4.1.8}$$

Now (4.1.8) implies that

$$D = T_{11}A + T_{12}B . \tag{4.1.9}$$

On the other hand, if we let $S = T^{-1}$ and partition S appropriately, then (4.1.8) implies that

$$\begin{bmatrix} A \\ B \end{bmatrix} = \begin{bmatrix} S_{11} & S_{12} \\ S_{21} & S_{22} \end{bmatrix} \begin{bmatrix} D \\ 0 \end{bmatrix} = \begin{bmatrix} S_{11}D \\ S_{21}D \end{bmatrix} . \tag{4.1.10}$$

Now (4.1.10) shows that D is a right divisor of both A and B, so that D satisfies (GCRD1). Further, if C is also a right divisor of A and B, then $A = A_1 C$, $B = B_1 C$ for some $A_1, B_1 \in \mathbf{M}(\mathbf{R})$. Then (4.1.9) shows that $D = (T_{11}A_1 + T_{12}B_1)C$. Hence, D is a left multiple of C, so that D satisfies (GCRD2). Thus, D is a g.c.r.d. of A, B and is of the form (4.1.5). Finally, suppose D_1 is any other g.c.r.d. of A, B. Then by Lemma 4.1.2 there exists a $U \in \mathbf{U}(\mathbf{R})$ such that $D_1 = UD$. Thus, $D_1 = UD = UT_{11}A + UT_{12}B$ is also of the form (4.1.5).

To complete the proof, we examine the case where F in (4.1.6) has fewer rows than columns. To be specific, suppose $A \in \mathbf{R}^{n \times m}$, $B \in \mathbf{R}^{l \times m}$ and that $n + l < m$. In this case, let $k = m - (n + l)$ and define

$$D = \begin{bmatrix} A \\ B \\ 0_{k \times m} \end{bmatrix} . \tag{4.1.11}$$

Then D is a right divisor of both A and B, since

$$A = [I_n \ 0 \ 0]\, D, \quad B = [0 \ I_l \ 0]\, D . \tag{4.1.12}$$

Moreover, D is of the form (4.1.5) since

$$D = \begin{bmatrix} I_n \\ 0 \\ 0 \end{bmatrix} A + \begin{bmatrix} 0 \\ I_l \\ 0 \end{bmatrix} B . \tag{4.1.13}$$

Now (4.1.12) and (4.1.13) are enough to show that D is a g.c.r.d. of A and B; the details are left to the reader. □

Corollary 4.1.4   *Suppose A, $B \in \mathbf{M}(\mathbf{R})$ have the same number of columns. Then A and B are right-coprime if and only if there exist X, $Y \in \mathbf{M}(\mathbf{R})$ such that*

$$XA + YB = I . \tag{4.1.14}$$

Proof. A and B are right-coprime if and only if I is a g.c.r.d. of A and B. Now apply the theorem. □

A useful way of interpreting Corollary 4.1.4 is the following: A, B are right-coprime if and only if the matrix $F = [A' \ B']'$ has a left inverse in $\mathbf{M}(\mathbf{R})$.

The relation (4.1.14) is called the *right Bezout identity*, or the *right Diophantine identity*. Though we have derived (4.1.14) as a corollary to Theorem 4.1.3, it turns out that (4.1.14) is the most useful characterization of right-coprimeness in subsequent applications.

Corollary 4.1.5   *Two matrices $A \in \mathbf{R}^{n \times m}$, $B \in \mathbf{R}^{l \times m}$ are right-coprime if and only if 1 is a g.c.d. of all $m \times m$ minors of the matrix $F = [A' \ B']'$.*

Remark 4.1.6   It is implicit in the above corollary that A, B cannot be right-coprime unless $n + l \geq m$.

Proof. "if" Suppose 1 is a g.c.d. of all $m \times m$ minors of F. Then all invariant factors of F are units (see Theorem B.2.6). Putting F into its Smith form, we see that there exist $U, V \in \mathbf{M}(\mathbf{R})$ such that

$$UFV = \begin{bmatrix} I \\ 0 \end{bmatrix}. \tag{4.1.15}$$

If U is partitioned in the obvious way, (4.1.15) implies that

$$(U_{11}A + U_{12}B)V = I, \tag{4.1.16}$$
$$U_{11}A + U_{12}B = V^{-1}, \tag{4.1.17}$$
$$VU_{11}A + VU_{12}B = I. \tag{4.1.18}$$

Since V is unimodular, (4.1.18) is of the form (4.1.14), with $X = VU_{11}$, $Y = VU_{12}$. Hence, A, B are right-coprime.

"only if" Suppose A, B are right-coprime. By (4.1.7) there exists a $T \in \mathbf{U}(\mathbf{R})$ such that

$$TF = \begin{bmatrix} D \\ 0 \end{bmatrix}, \tag{4.1.19}$$

where D is upper triangular and is a g.c.r.d. of A, B. Since A, B are right-coprime, D is unimodular. Hence,

$$TFD^{-1} = \begin{bmatrix} I \\ 0 \end{bmatrix}. \tag{4.1.20}$$

Since T and D^{-1} are both unimodular, the right side of (4.1.15) is a Smith form of F. By Theorem B.2.6, it follows that 1 is a g.c.d. of all $m \times m$ minors of F. □

Corollary 4.1.7 *Suppose A, $B \in \mathbf{M}(\mathbf{R})$ have the same number of columns, and let $F = [A'\ B']'$. Then A, B are right-coprime if and only if there exists a $U \in \mathbf{U}(\mathbf{R})$ of the form $U = [F\ G]$.*

Remarks 4.1.8 To put it into words, Corollary 4.1.7 states that A and B are right-coprime if and only if there exists a unimodular matrix containing $[A'\ B']$ among its columns, or equivalently, if and only if the matrix F can be "complemented" by appropriate columns to form a unimodular matrix. For a generalization of this property to more general situations, see Section 8.1.

Proof. "if" Suppose there exists a $U \in \mathbf{U}(\mathbf{R})$ of the form $[F\ G]$, and let $V = U^{-1}$. Partitioning V appropriately and expanding $VU = I$ gives

$$V_{11}A + V_{12}B = I. \tag{4.1.21}$$

Hence, A, B are right-coprime.

"only if" Suppose A, B are right-coprime. By Corollary 4.1.5, there exist T, $V \in \mathbf{U}(\mathbf{R})$ such that

$$T \begin{bmatrix} A \\ B \end{bmatrix} V = \begin{bmatrix} I \\ 0 \end{bmatrix} , \tag{4.1.22}$$

or equivalently,

$$T \begin{bmatrix} A \\ B \end{bmatrix} = \begin{bmatrix} V^{-1} \\ 0 \end{bmatrix} . \tag{4.1.23}$$

Let $W = T^{-1}$ and partition T, W appropriately. Then

$$\begin{bmatrix} T_{11} & T_{12} \\ T_{21} & T_{22} \end{bmatrix} \begin{bmatrix} A & W_{12} \\ B & W_{22} \end{bmatrix} = \begin{bmatrix} V^{-1} & 0 \\ 0 & I \end{bmatrix} , \tag{4.1.24}$$

where we use the fact that $TW = I$. As the matrix on the right side of (4.1.24) is unimodular, so are both the matrices on the left side. If we define

$$U = \begin{bmatrix} A & W_{12} \\ B & W_{22} \end{bmatrix} , \tag{4.1.25}$$

then the proof is complete.   □

Suppose A, $B \in \mathbf{M}(\mathbf{R})$ are right-coprime. Then, by Corollary 4.1.4, there exist X, $Y \in \mathbf{M}(\mathbf{R})$ such that $XA + YB = I$. Lemma 4.1.9 below characterizes *all* solutions of the Equation (4.1.14). It turns out that this is an important part of the controller synthesis theory to be developed in Chapter 5.

Lemma 4.1.9   *Suppose A, $B \in \mathbf{M}(\mathbf{R})$ are right-coprime, and select $U \in \mathbf{U}(\mathbf{R})$ of the form in Corollary 4.1.7 (i.e., such that U contains $[A' \ B']'$ as its first columns). Then the set of X, $Y \in \mathbf{M}(\mathbf{R})$ that satisfy*

$$XA + YB = I , \tag{4.1.26}$$

is given by

$$[X \ \ Y] = [I \ \ R]U^{-1}, \quad R \in \mathbf{M}(\mathbf{R}) . \tag{4.1.27}$$

Proof. It is necessary to show that (i) every X, Y of the form (4.1.27) satisfies (4.1.26), and (ii) every X, Y that satisfies (4.1.26) is of the form (4.1.27) for some $R \in \mathbf{M}(\mathbf{R})$.

To prove (i), observe that $U^{-1}U = I$. Hence,

$$XA + YB = [X \ \ Y] \begin{bmatrix} A \\ B \end{bmatrix}$$
$$= [I \ \ R]U^{-1} \begin{bmatrix} A \\ B \end{bmatrix}$$
$$= [I \ \ R] \begin{bmatrix} I \\ 0 \end{bmatrix} = I , \tag{4.1.28}$$

so that X, Y satisfy (4.1.26). To prove (ii), suppose X, Y satisfy (4.1.26) and define $R = [X \ Y]G$. Then

$$[X \ Y]U = [X \ Y] \begin{bmatrix} A \\ & G \\ B \end{bmatrix} = [I \ R] \qquad (4.1.29)$$

from which (4.1.27) follows readily.   □

The concept of right-coprimeness can be readily extended to more than two matrices. Suppose $A_1, \cdots, A_p \in \mathbf{M}(\mathbf{R})$ all have the same number of columns. Then a square matrix D is a *g.c.r.d.* of $A_1, \cdots, A_p$ if

(i) D is a right divisor of A_i for all i, and

(ii) If C is a right divisor of A_i for all i, then D is a left multiple of C.

The matrices $A_1, \cdots, A_p$ are *right-coprime* if every g.c.r.d. of $A_1, \cdots, A_p$ is unimodular.

The following facts are straight-forward extensions of the corresponding results for the case of two matrices; the proofs are left to the reader.

Fact 4.1.10   Suppose $A_1, \cdots, A_p \in \mathbf{M}(\mathbf{R})$ all have the same number of columns. Then they have a g.c.r.d. Moreover, if D is *any* g.c.r.d. of $A_1, \cdots, A_p$, then there exist $X_1, \cdots, X_p \in \mathbf{M}(\mathbf{R})$ such that

$$D = \sum_{i=1}^{p} X_i A_i \ . \qquad (4.1.30)$$

Fact 4.1.11   Suppose $A_1, \cdots, A_p \in \mathbf{M}(\mathbf{R})$ all have the same number of columns, and define

$$F = \begin{bmatrix} A_1 \\ A_2 \\ \vdots \\ A_p \end{bmatrix} . \qquad (4.1.31)$$

Then the following statements are equivalent:

(i) $A_1, \cdots, A_p$ are right-coprime.

(ii) F has a left inverse in $\mathbf{M}(\mathbf{R})$.

(iii) F has at least as many rows as columns, and 1 is a g.c.d. of all the largest-size minors of F.

(iv) There exists a $U \in \mathbf{U}(\mathbf{R})$ of the form $U = [F \ ; \ G]$.

Now suppose $A_1, \cdots, A_p$ are right-coprime. Then the set of all left inverses in $\mathbf{M(R)}$ of F is given by

$$\{X : X = [I\ R]U^{-1}, \quad R \in \mathbf{M(R)}\}, \tag{4.1.32}$$

where U is any unimodular matrix of the form $[F\ G]$.

The next definition introduces a concept that is central to this book.

Definition 4.1.12   Suppose $P \in \mathbf{M(F)}$. An ordered pair (N, D) where $N, D \in \mathbf{M(R)}$ is a *right-coprime factorization (r.c.f.) of P* if
(RCF1) D is square and $|D| \neq 0$,
(RCF2) $P = ND^{-1}$, and
(RCF3) N and D are right-coprime.

Theorem 4.1.13   Suppose $P \in \mathbf{M(F)}$. Then it has an r.c.f. (N, D). Further (NU, DU) is also an r.c.f. of P for every $U \in \mathbf{U(R)}$. Finally, suppose (N_1, D_1) is another r.c.f. of P. Then there exists a $U \in \mathbf{U(R)}$ such that $N_1 = NU, D_1 = DU$.

Remarks 4.1.14   Theorem 4.1.13 states that every $P \in \mathbf{M(F)}$ has an r.c.f., and that the r.c.f. is unique except for the possibility of multiplying the "numerator" and "denominator" matrices on the right by a unimodular matrix.

Proof. (There are many ways to prove that every $P \in \mathbf{M(F)}$ has an r.c.f. One way is given below. See the proof of Theorem 4.1.15 for another method of constructing r.c.f.'s.) Write every element p_{ij} as a_{ij}/b_{ij}, where $a_{ij} \in \mathbf{R}, b_{ij} \in \mathbf{R} \setminus \{0\}$. Let b denote a least common multiple of all the b_{ij}'s, and let $F = bP$. Then $F \in \mathbf{M(R)}$ and $P = F(bI)^{-1}$. In accordance with Theorem 4.1.3, let E be a g.c.r.d. of F and bI, and define $N, D \in \mathbf{M(R)}$ by $F = NE, bI = DE$. Then, as in (4.1.5), there exist $X, Y \in \mathbf{M(R)}$ such that

$$XF + Yb = E. \tag{4.1.33}$$

Since $|D| \cdot |E| = |bI| \neq 0$, it follows that $|D| \neq 0, |E| \neq 0$. Hence, we can multiply both sides of (4.1.33) by E^{-1}, which gives

$$XN + YD = I, \tag{4.1.34}$$

showing that N and D are right-coprime. Finally, it is immediate that $ND^{-1} = (NE)(DE)^{-1} = P$. Hence, (N, D) is an r.c.f. of P.

It is left to the reader to verify that (NU, DU) is also an r.c.f. of P whenever $U \in \mathbf{U(R)}$.

To prove the final assertion, let (N_1, D_1) be another r.c.f. of P. Then, by definition, $|D_1| \neq 0, P = N_1 D_1^{-1}$ and there exist $X_1, Y_1 \in \mathbf{M(R)}$ such that

$$X_1 N_1 + Y_1 D_1 = I. \tag{4.1.35}$$

Multiplying both sides by D_1^{-1} gives

$$X_1 N_1 D_1^{-1} + Y_1 = D_1^{-1} . \tag{4.1.36}$$

Multiplying both sides by D, and noting that $N_1 D_1^{-1} = N D^{-1}$, gives

$$D_1^{-1} D = X_1 N + Y_1 D \in \mathbf{M}(\mathbf{R}) . \tag{4.1.37}$$

By symmetrical reasoning, we also get $D^{-1} D_1 := U \in \mathbf{M}(\mathbf{R})$. Since U has an inverse in $\mathbf{M}(\mathbf{R})$, it is unimodular. Further, $D_1 = DU$ and $N_1 = P D_1 = PDU = NU$.   □

It is left to the reader to state and prove the "left" analogs of the results presented up to now. These results are used freely in the sequel, without comment. As a mnemonic aid, a tilde "~" is used to denote "left" quantities. For example, (N, D) is used for an r.c.f. of $P \in \mathbf{M}(\mathbf{F})$ and $(\tilde{D}, \tilde{N})$ for an l.c.f. of P. Note that the order of the "denominator" and "numerator" matrices is interchanged in the latter case. This is to reinforce the point that if (N, D) is an r.c.f. of P, then $P = N D^{-1}$, whereas if $(\tilde{D}, \tilde{N})$ is an l.c.f. of P, then $P = \tilde{D}^{-1} \tilde{N}$.

Up to now, it has been shown that every $P \in \mathbf{M}(\mathbf{F})$ has both an r.c.f. and an l.c.f. The next question to be answered is whether there is any relationship between the two.

Theorem 4.1.15   Suppose $P \in \mathbf{M}(\mathbf{F})$ and that $(N, D), (\tilde{D}, \tilde{N})$ are an r.c.f. and an l.c.f. of P. Then $N, \tilde{N}$ are equivalent, and the nonunit invariant factors of $D, \tilde{D}$ are the same. In particular, $|D|$ and $|\tilde{D}|$ are associates.[2]

Proof. Suppose to be specific that P is of order $n \times m$ and has rank r. The discussion below applies to the case where $n \leq m$; the case $n > m$ is entirely similar and is left to the reader. For convenience, let l denote $m - n$.

Putting P in Smith-McMillan form (Theorem B.2.9), we see that there exist $U, V \in \mathbf{U}(\mathbf{R})$ such that

$$UPV = \begin{bmatrix} a_1/b_1 & & & & \\ & \cdot & & & 0 \\ & & \cdot & & 0_{n \times l} \\ & & \cdot & a_r/b_r & \\ 0 & & & & \end{bmatrix} , \tag{4.1.38}$$

where a_i, b_i are coprime for all i. Now define $A = \text{Diag}\{a_1, \cdots, a_r\}$, $B = \text{Diag}\{b_1, \cdots, b_r\}$, and let

$$N_1 = U^{-1} \begin{bmatrix} A & 0 & \\ & & 0_{n \times l} \\ 0 & 0 & \end{bmatrix} \tag{4.1.39}$$

$$D_1 = V \begin{bmatrix} B & 0 \\ 0 & I_{m-r} \end{bmatrix} . \tag{4.1.40}$$

[2]See Section B.2 for the definition of equivalence.

Then $|D_1| = \prod_{i=1}^{r} b_i \neq 0$, and $P = N_1 D_1^{-1}$. Also, since a_i, b_i are coprime for all i, there exist elements x_i, $y_i \in \mathbf{R}$ such that $x_i a_i + y_i b_i = 1$ for all i. Now define the matrices $X, Y \in \mathbf{R}^{n \times n}$ by $X =$ Diag $\{x_1, \cdots, x_r, 0, \cdots, 0\}$, $Y =$ Diag $\{y_1, \cdots, y_r, 1, \cdots, 1\}$, and let

$$S = \begin{bmatrix} X \\ 0_{l \times n} \end{bmatrix}, \tag{4.1.41}$$

then it follows that

$$SN_1 + YD_1 = I. \tag{4.1.42}$$

Thus, (N_1, D_1) is an r.c.f. of P.

Using entirely parallel reasoning, one can show that $(\tilde{D}_1, \tilde{N}_1)$ is an l.c.f. of P, where

$$\tilde{N}_1 = \begin{bmatrix} A & 0 & \\ & & 0_{n \times l} \\ 0 & 0 & \end{bmatrix} V, \tag{4.1.43}$$

$$\tilde{D}_1 = \begin{bmatrix} B & 0 \\ 0 & I_{n-r} \end{bmatrix} U^{-1}. \tag{4.1.44}$$

It is clear from (4.1.39) and (4.1.43) that N_1 and $\tilde{N}_1$ are equivalent matrices, as they both have the same invariant factors (see Corollary B.2.8). Similarly, (4.1.40) and (4.1.44) show that D_1 and $\tilde{D}_1$ have the same nonunit invariant factors. Thus, we have established the theorem for the *particular* r.c.f. (N_1, D_1) and l.c.f. $(\tilde{D}_1, \tilde{N}_1)$ of P.

To complete the proof, let (N, D), $(\tilde{D}, \tilde{N})$ be *any* r.c.f. and l.c.f. of P. Then by Theorem 4.1.13 and its left analog, there exist unimodular matrices $W \in \mathbf{R}^{m \times m}$, $Z \in \mathbf{R}^{n \times n}$ such that

$$N = N_1 W, \quad \tilde{N} = Z \tilde{N}_1, \tag{4.1.45}$$
$$D = D_1 W, \quad \tilde{D} = Z \tilde{D}_1. \tag{4.1.46}$$

Now (4.1.45) shows that N and N_1 are equivalent, and that $\tilde{N}$ and $\tilde{N}_1$ are equivalent. Since equivalence of matrices is transitive, it follows that N and $\tilde{N}$ are also equivalent. Similarly, (4.1.46) shows that D and $\tilde{D}$ have the same nonunit invariant factors. Finally, the fact that $|D|$ and $|\tilde{D}|$ are associates is a direct consequence of the preceding sentence. □

We conclude this section by introducing yet another type of useful factorization.

Theorem 4.1.16   Suppose $P \in \mathbf{M}(\mathbf{F})$, and let (N, D), $(\tilde{D}, \tilde{N})$ be any r.c.f. and l.c.f. of P. Suppose $X, Y \in \mathbf{M}(\mathbf{R})$ satisfy

$$XN + YD = I. \tag{4.1.47}$$

Then there exist $\tilde{X}, \tilde{Y} \in \mathbf{M}(\mathbf{R})$ such that

$$\begin{bmatrix} Y & X \\ -\tilde{N} & \tilde{D} \end{bmatrix} \begin{bmatrix} D & -\tilde{X} \\ N & \tilde{Y} \end{bmatrix} = I \ . \tag{4.1.48}$$

Remark 4.1.17   The ordered pair of matrices in (4.1.47) is referred to as a *doubly coprime factorization* of P.

Proof. Given $N, D, \tilde{N}, \tilde{D}$, there exist $\tilde{X}_1, \tilde{Y}_1 \in \mathbf{M}(\mathbf{R})$ such that

$$\tilde{N}\tilde{X}_1 + \tilde{D}\tilde{Y}_1 = I \ , \tag{4.1.49}$$

since $\tilde{N}, \tilde{D}$ are left-coprime. Define

$$E = \begin{bmatrix} Y & X \\ -\tilde{N} & \tilde{D} \end{bmatrix} . \tag{4.1.50}$$

Then

$$E = \begin{bmatrix} D & -\tilde{X}_1 \\ N & \tilde{Y}_1 \end{bmatrix} = \begin{bmatrix} I & \Delta \\ 0 & I \end{bmatrix} , \tag{4.1.51}$$

where $\Delta = -Y\tilde{X}_1 + X\tilde{Y}_1$. (In (4.1.51) we use the fact that $\tilde{N}D = \tilde{D}N$, since $\tilde{D}^{-1}\tilde{N} = ND^{-1}$.) Since the matrix on the right side of (4.1.51) is unimodular, so is E. Moreover, from (4.1.51),

$$E^{-1} = \begin{bmatrix} D & -\tilde{X}_1 \\ N & \tilde{Y}_1 \end{bmatrix} \begin{bmatrix} I & \Delta \\ 0 & I \end{bmatrix}^{-1} = \begin{bmatrix} D & -(\tilde{X}_1 + D\Delta) \\ N & \tilde{Y}_1 + N\Delta \end{bmatrix} . \tag{4.1.52}$$

Thus, (4.1.48) is satisfied with, $\tilde{X} = \tilde{X}_1 + D\Delta, \tilde{Y} = \tilde{Y}_1 + N\Delta.$ □

The above proof makes it clear that, once $N, D, \tilde{N}, \tilde{D}$ have been selected, every X, Y such that $XN + YD = I$ determines a unique $\tilde{X}, \tilde{Y}$ such that (4.1.47) holds, and vice-versa. The proof also demonstrates the following result.

Corollary 4.1.18   *Suppose $P \in \mathbf{M}(\mathbf{F})$, and that $(N, D), (\tilde{D}, \tilde{N})$ are a r.c.f. and l.c.f. of P. Then for every $X, Y, \tilde{X}, \tilde{Y} \in \mathbf{M}(\mathbf{R})$ such that $XN + YD = I, \tilde{N}\tilde{X} + \tilde{D}\tilde{Y} = I$. the matrices*

$$U_1 = \begin{bmatrix} Y & X \\ -\tilde{N} & \tilde{D} \end{bmatrix}, \quad U_2 = \begin{bmatrix} D & -\tilde{X} \\ N & \tilde{Y} \end{bmatrix}, \tag{4.1.53}$$

are unimodular. Moreover, U_1^{-1} is a complementation of $[D'\ N']'$, in that U_1^{-1} is of the form

$$U_1^{-1} = \begin{bmatrix} D \\ N \end{bmatrix} G,$$ \hfill (4.1.54)

for some $G \in \mathbf{M}(\mathbf{R})$. Similarly, U_2^{-1} is a complementation of the matrix $[-\tilde{N}\ \ \tilde{D}]$ in that U_2^{-1} is of the form

$$U_2^{-1} = \begin{bmatrix} H \\ -\tilde{N} & \tilde{D} \end{bmatrix},$$ \hfill (4.1.55)

for some $H \in \mathbf{M}(\mathbf{R})$.

PROBLEMS

4.1.1. As an illustration of Lemma 4.1.2, let $\mathbf{R}$ be the ring $\mathbb{R}[s]$ of polynomials with real coefficients. Suppose

$$A = \begin{bmatrix} 1 & s \\ s^2 & s^3 \end{bmatrix}, \quad B = [s+1\ \ s^2+s].$$

(i) Show that both

$$D = \begin{bmatrix} 1 & s \\ 0 & 0 \end{bmatrix}, \quad D_1 = \begin{bmatrix} 1 & s \\ s & s^2 \end{bmatrix},$$

are g.c.r.d.'s of A and B.

(ii) Verify that $D_1 = VD$, where

$$V = \begin{bmatrix} 1 & s \\ s & 1 \end{bmatrix},$$

and that V is *not* unimodular over $\mathbb{R}[s]$.

(iii) Find a unimodular matrix U such that $D_1 = UD$.

4.1.2. Suppose $A, B \in \mathbf{M}(\mathbf{R})$ have the same number of columns, and let $F = [A'\ B']'$. Show that every g.c.r.d. of A, B is nonsingular if and only if F has full column rank. (Hint: See (4.1.7).)

4.1.3. Suppose $A, B \in \mathbf{M}(\mathbf{R})$ are right-coprime. Suppose A is square and let $a = |A|$. Finally, let b denote the smallest invariant factor of B. Show that a and b are coprime.

4.1.4. Prove the following generalization of Problem 4.1.3: Suppose $A \in \mathbf{R}^{n \times n}, B \in \mathbf{R}^{m \times n}$ are right-coprime. Let $a_1, \cdots, a_n$ denote the invariant factors of A, and let $b_1, \cdots, b_n (b_1, \cdots, b_m$ if $m < n)$ denote the invariant factors of B. Then a_n, b_1 are coprime, a_{n-1}, b_2 are coprime, $\ldots$, a_1, b_n are coprime. (If $m < n$, then $a_m, a_{m-1}, \cdots, a_1$ are all units.)

4.1.5. Prove, or disprove by means of an example: Suppose A and B are square and right-coprime. Then $|A|$ and $|B|$ are coprime.

4.1.6. Prove, or disprove by means of an example: Suppose $A, B \in \mathbf{M(R)}$ are square and $|A|, |B|$ are coprime. Then A, B are right- as well as left-coprime.

4.1.7. Show that $A, B \in \mathbf{M(R)}$ are right-coprime if and only if A', B' are left-coprime.

4.1.8. Prove Fact 4.1.10.

4.1.9. Prove Fact 4.1.11.

4.1.10. Suppose $A, B \in \mathbf{M(R)}$ are right-coprime. Show that $A + RB, B$ are right-coprime for all $R \in \mathbf{M(R)}$. State and prove the left analog of this result.

4.1.11. Suppose $G \in \mathbf{M(F)}$, $H \in \mathbf{M(R)}$ have the same order, and let (N, D) be an r.c.f. of G. Show that $(N + HD, D)$ is an r.c.f. of $G + H$. State and prove the left analog of this result. (Hint: See Problem 4.1.10.)

4.1.12. This problem is a generalization of Problem 4.1.11. Suppose $G \in \mathbf{M(F)}$, $H \in \mathbf{M(R)}$ and let $N, D, \tilde{N}, \tilde{D}, X, Y, \tilde{X}, \tilde{Y} \in \mathbf{M(R)}$ be a doubly coprime factorization of G. Thus, (4.1.48) holds, and in addition $G = ND^{-1} = \tilde{D}^{-1}\tilde{N}$. Show that a doubly coprime factorization of $G + H$ can be obtained by the following replacements:

$$N \leftarrow N + HD, \; Y \leftarrow Y - XH, \; \tilde{N} \leftarrow \tilde{N} + \tilde{D}H, \; \tilde{Y} \leftarrow \tilde{Y} - H\tilde{X} \, .$$

4.2   COPRIME FACTORIZATIONS OVER S

In this section, a method is presented for obtaining left- and right-coprime factorizations of a system transfer matrix from its state-space description. This method makes it easy to apply the subsequently developed synthesis theory to systems described in state-space form.

Consider a system described by the equations

$$\dot{x}(t) = Ax(t) + Bu(t) \, , \tag{4.2.1}$$
$$y(t) = Cx(t) + Eu(t) \, , \tag{4.2.2}$$

where A, B, C, E are constant matrices of compatible dimensions. The transfer matrix of this system is

$$P(s) = C(sI - A)^{-1}B + E \, . \tag{4.2.3}$$

The objective is to derive a doubly coprime factorization of P. One such factorization is given in Theorem 4.2.1 below.

Theorem 4.2.1   Given the system (4.2.1)–(4.2.2), suppose the pairs (A, B), (A, C) are stabilizable and detectable, respectively. Select constant matrices K and F such that the matrices $A_0 := A -$

BK, $\tilde{A}_0 := A - FC$ are both Hurwitz.[3] Then $P = N_p D_p^{-1} = \tilde{D}_p^{-1} \tilde{N}_p$ and

$$\begin{bmatrix} Y & X \\ -\tilde{N}_p & \tilde{D}_p \end{bmatrix} \begin{bmatrix} D_p & -\tilde{X} \\ N_p & \tilde{Y} \end{bmatrix} = I \ . \tag{4.2.4}$$

where the various matrices are defined as follows:

$$\begin{aligned}
\tilde{N}_p &= C(sI - \tilde{A}_0)^{-1}(B - FE) + E \ , \\
\tilde{D}_p &= I - C(sI - \tilde{A}_0)^{-1}F \ , \\
N_p &= (C - EK)(sI - A_0)^{-1}B + E \ , \\
D_p &= I - K(sI - A_0)^{-1}B \ , \\
X &= K(sI - \tilde{A}_0)^{-1}F \ , \\
Y &= I + K(sI - \tilde{A}_0)^{-1}(B - FE) \ , \\
\tilde{X} &= K(sI - A_0)^{-1}F \ , \\
\tilde{Y} &= I + (C - EK)(sI - A_0)^{-1}F \ .
\end{aligned} \tag{4.2.5}$$

As shown in Problem 4.1.12, if we can find a doubly coprime factorization for the transfer matrix

$$G(s) = P(s) - E = C(sI - A)^{-1}B \ , \tag{4.2.6}$$

then it is a simple matter to find a corresponding doubly coprime factorization for $P = G + E$. Hence, the proof of Theorem 4.2.1 is straight-forward once the following lemma is established.

Lemma 4.2.2   *Let all symbols be as in Theorem 4.2.1, and define*

$$G(s) = C(sI - A)^{-1}B \ . \tag{4.2.7}$$

Then a doubly coprime factorization of G is given by the matrices

$$\begin{aligned}
\tilde{N}_g &= C(sI - \tilde{A}_0)^{-1}B \ , \\
\tilde{D}_g &= I - C(sI - \tilde{A}_0)^{-1}F \ , \\
N_g &= C(sI - A_0)^{-1}B \ , \\
D_g &= I - K(sI - A_0)^{-1}B \ , \\
X_g &= K(sI - \tilde{A}_0)^{-1}F \ , \\
Y_g &= I + K(sI - \tilde{A}_0)^{-1}B \ , \\
\tilde{X}_g &= K(sI - A_0)^{-1}F \ , \\
\tilde{Y}_g &= I + C(sI - A_0)^{-1}F \ .
\end{aligned} \tag{4.2.8}$$

[3]A matrix is *Hurwitz* if all its eigenvalues have negative real parts.

Proof. It is first shown that $G = N_g D_g^{-1} = \tilde{D}_g^{-1} \tilde{N}_g$. This part of the proof uses the matrix identity $T(I + UT)^{-1} = (I + TU)^{-1} T$. Now

$$
\begin{aligned}
G &= C(sI - A)^{-1} B \\
  &= C(sI - A_0 - BK)^{-1} B \\
  &= C[(I - BK(sI - A_0)^{-1})(sI - A_0)]^{-1} B \\
  &= C(sI - A_0)^{-1}[I - BK(sI - A_0)^{-1}]^{-1} B \\
  &= C(sI - A_0)^{-1} B[I - K(sI - A_0)^{-1} B]^{-1} \\
  &= N_g D_g^{-1} .
\end{aligned}
\tag{4.2.9}
$$

The proof that $G = \tilde{D}_g^{-1} \tilde{N}_g$, is entirely similar.

The next part of the proof shows that the Bezout identities are satisfied. We have

$$
\begin{aligned}
\tilde{N}_g \tilde{X}_g + \tilde{D}_g \tilde{Y}_g &= C(sI - \tilde{A}_0)^{-1} BK(sI - A_0)^{-1} F \\
&\quad + [I - C(sI - \tilde{A}_0)^{-1} F][I + C(sI - A_0)^{-1} F] \\
&= C(sI - \tilde{A}_0)^{-1} BK(sI - A_0)^{-1} F \\
&\quad + I - C(sI - \tilde{A}_0)^{-1} F + C(sI - A_0)^{-1} F \\
&\quad - C(sI - \tilde{A}_0)^{-1} FC(sI - A_0)^{-1} F .
\end{aligned}
\tag{4.2.10}
$$

To show that $\tilde{N}_g \tilde{X}_g + \tilde{D}_g \tilde{Y}_g = I$, it suffices to show that

$$
C(sI - \tilde{A}_0)^{-1}[BK - (sI - A_0) + sI - \tilde{A}_0 - FC](sI - A_0)^{-1} F = 0 .
\tag{4.2.11}
$$

But this relation is immediate, since the matrix inside the brackets is zero. The remaining equations follow in the same manner. $\qquad\square$

The proof of Theorem 4.2.1 based on Lemma 4.2.2 is left to the reader.

4.3   BICOPRIME FACTORIZATIONS, CHARACTERISTIC DETERMINANTS

The objectives of this section are two-fold: (i) to introduce the concept of bicoprime factorizations, and (ii) to define characteristic determinants and McMillan degree, and to relate these to bicoprime factorizations. Throughout this section, **R** denotes a principal ideal domain and **F** the field of fractions associated with **R**.

Definition 4.3.1   Suppose $F \in \mathbf{M}(\mathbf{F})$. A quadruple $(N_r, D, N_l, K) \in \mathbf{M}(\mathbf{R})$ is a *bicoprime factorizarion of F* if (i) $|D| \neq 0$ and $F = N_r D^{-1} N_l + K$, (ii) N_r, and D are right-coprime, and (iii) N_l and D are left-coprime.

Note that the term right-left-coprime factorization is also used by other authors.

There are at least two good reasons for studying bicoprime factorizations: 1) In the study of feedback systems, such as in Figure 1.2, it turns out that, starting from coprime factorizations for the plant and compensator, one is led naturally to a *bicoprime* (and not coprime) factorization of the resulting closed-loop transfer matrix (see Section 5.1). 2) In the case where the ring $\mathbf{R}$ is $\mathbb{R}[s]$, if a system is described in state-space form by

$$\dot{x}(t) = Ax(t) + Bu(t), \quad y(t) = Cx(t) + Eu(t), \tag{4.3.1}$$

then its transfer matrix $H(s)$ equals $C(sI - A)^{-1}B + E$; further, if the description (4.3.1) is *minimal*, then $(C, sI - A, B, E)$ is a bicoprime factorization of H.

It is easy to see that if $N_l = I$, $K = 0$, (resp. $N_r = I$, $K = 0$), then a bicoprime factorization becomes an r.c.f. (resp. an l.c.f.).

Definition 4.3.2 Suppose $F \in \mathbf{M(F)}$. A *characteristic determinant* of F is an element $\phi \in \mathbf{R}$ such that

(i) the product of ϕ and every minor of F belongs to $\mathbf{R}$, and

(ii) if $\psi \in \mathbf{R}$ also satisfies (i) above (i.e., the product of ψ and every minor of F belongs to $\mathbf{R}$), then ϕ divides ψ.

An equivalent, and more transparent, definition of a characteristic determinant is the following: Consider all minors of F, and express them as reduced-form fractions. Then ϕ is a least common multiple of all the denominators of the various minors of F. Since a least common multiple is only unique to within multiplication by a unit of $\mathbf{R}$, the same is true of the characteristic determinant.

Note that in the case where $\mathbf{R} = \mathbb{R}[s]$ and $\mathbf{F} = \mathbb{R}(s)$, it is customary to refer to the characteristic determinant as the characteristic polynomial.

Example 4.3.3 Let $\mathbf{R} = \mathbb{R}[s]$, $\mathbf{F} = \mathbb{R}(s)$, and suppose

$$F = \begin{bmatrix} \dfrac{1}{s-1} & \dfrac{2}{s} \\ \dfrac{-s}{s+1} & \dfrac{s}{s^2-1} \end{bmatrix}. \tag{4.3.2}$$

Then F has five minors (four 1×1 and one 2×2). The monic l.c.m. of all the denominators of these minors is $\phi(s) = s(s+1)(s-1)^2$. Hence, the set of *all* characteristic polynomials of F is $a\,\phi(s)$, where a is any nonzero real number.

It is obvious from the definition that $F \in \mathbf{M(F)}$ actually belongs to $\mathbf{M(R)}$ if and only if its characteristic determinant is a unit.

The next theorem is the main result of this section.

Theorem 4.3.4 Suppose $F \in \mathbf{M(F)}$, and that (N_r, D, N_l, K) is a bicoprime factorization of F. Then $|D|$ is a characteristic determinant of F.

The proof requires a few lemmas, which are presented next. In the interests of brevity, let $\phi(A)$ denote a characteristic determinant of $A \in \mathbf{M}(\mathbf{F})$.

Lemma 4.3.5   *Suppose $A, B \in \mathbf{M}(\mathbf{F})$ and $A - B \in \mathbf{M}(\mathbf{R})$. Then $\phi(A)$ and $\phi(B)$ are associates.*

Proof. Suppose $A = B + C$ where $C \in \mathbf{M}(\mathbf{R})$. It is first shown that $\phi(A)$ divides $\phi(B)$. Consider a typical (say $n \times n$) minor of A, of the form

$$\begin{vmatrix} a_{i_1 j_1} & \cdots & a_{i_1 j_n} \\ \vdots & & \vdots \\ a_{i_n j_1} & \cdots & a_{i_n j_n} \end{vmatrix} = \begin{vmatrix} b_{i_1 j_1} + c_{i_1 j_1} & \cdots & b_{i_1 j_n} + c_{i_1 j_n} \\ \vdots & & \vdots \\ b_{i_n j_1} + c_{i_n j_1} & \cdots & b_{i_n j_n} + c_{i_n j_n} \end{vmatrix}. \tag{4.3.3}$$

Since the determinant is a multilinear function, the minor in (4.3.3) can be expanded as a sum of 2^n determinants, namely

$$\begin{vmatrix} b_{i_1 j_1} & \cdots & b_{i_1 j_n} \\ \vdots & & \vdots \\ b_{i_n j_1} & \cdots & b_{i_n j_n} \end{vmatrix} + \begin{vmatrix} c_{i_1 j_1} & b_{i_1 j_2} & \cdots & b_{i_1 j_n} \\ \vdots & \vdots & & \vdots \\ c_{i_n j_1} & b_{i_n j_2} & \cdots & b_{i_n j_n} \end{vmatrix} + \cdots + \begin{vmatrix} c_{i_1 j_1} & \cdots & c_{i_1 j_n} \\ \vdots & & \vdots \\ c_{i_n j_1} & \cdots & c_{i_n j_n} \end{vmatrix}. \tag{4.3.4}$$

Moreover, each of the $2^n - 1$ minors involving some elements of C can be further expanded, using Laplace's expansion, into a sum of products of minors of B and of C. Since $C \in \mathbf{M}(\mathbf{R})$, every minor of C belongs to $\mathbf{R}$. Since the product of $\phi(B)$ and every minor of B belongs to $\mathbf{R}$, we conclude that the product of $\phi(B)$ and the sum in (4.3.4) belongs to $\mathbf{R}$. Since this is true of *every* minor of B, it follows that $\phi(A)$ divides $\phi(B)$, by condition (ii) of Definition 4.3.2.

Next, write $B = A + (-C)$ and note that $-C \in \mathbf{M}(\mathbf{R})$. By the same reasoning as in the preceding paragraph, $\phi(B)$ divides $\phi(A)$. Hence, $\phi(A)$ and $\phi(B)$ are associates.   □

Lemma 4.3.6   *Suppose $X, Y \in \mathbf{M}(\mathbf{R})$, $B \in \mathbf{M}(\mathbf{F})$, and let $A = XBY$. Then $\phi(A)$ divides $\phi(B)$.*

Proof. To be specific, suppose X, B, Y have dimensions $k \times l, l \times m, m \times n$, respectively. Let $r \leq \min\{k, l, m, n\}$ and suppose $J \in S(r, k)$, $K \in S(r, n)$. By the Binet-Cauchy formula (Fact B.1.5), the minor a_{JK} is given by

$$a_{JK} = \sum_{K_1 \in S(r,l)} \sum_{K_2 \in S(r,m)} x_{JK_1} b_{K_1 K_2} y_{K_2 K}. \tag{4.3.5}$$

Hence,

$$a_{JK} \cdot \phi(B) = \sum_{K_1} \sum_{K_2} x_{JK_1} \cdot [b_{K_1 K_2} \phi(B)] \cdot y_{K_2 K}. \tag{4.3.6}$$

Since every term in the summation on the right side of (4.3.6) belongs to $\mathbf{R}$, so does $a_{JK} \cdot \phi(B)$ for all J, K. Hence, by condition (ii) of Definition 4.3.2, $\phi(A)$ divides $\phi(B)$.   □

Lemma 4.3.7 *Suppose $P, Q, R \in \mathbf{M}(\mathbf{R})$, $|Q| \neq 0$, and let $A = PQ^{-1}R \in \mathbf{M}(\mathbf{F})$. Then $\phi(A)$ divides $|Q|$.*

Proof. By Fact B.1.8, $(Q^{-1})_{JK} \cdot |Q|$ equals $\pm$ an appropriate minor of Q, and hence belongs to $\mathbf{R}$, for all J, K. Hence, $\phi(Q^{-1})$ divides $|Q|$. By Lemma 4.3.6, $\phi(A)$ divides $\phi(Q^{-1})$. The conclusion follows. □

Lemma 4.3.8 *Suppose $Q \in \mathbf{M}(\mathbf{R})$ is square and $|Q| \neq 0$. Then $\phi(Q^{-1}) = |Q|$. .*

Proof. As in Lemma 4.3.7, it is immediate that $\phi(Q^{-1})$ divides $|Q|$. To prove the converse, note that $|Q^{-1}| = 1/|Q|$. Hence, any l.c.m. of the denominators of all minors of Q^{-1} must, in particular, be a multiple of $|Q|$. □

Proof of Theorem 4.3.4. It is first shown that $\phi(F)$ divides $|D|$. We know that $F = N_r D^{-1} N_l + K$, and since $K \in \mathbf{M}(\mathbf{R})$ it follows from Lemma 4.3.5 that $\phi(F) = \phi(F - K) = \phi(N_r D^{-1} N_l)$. By Lemma 4.3.7, the latter divides $|D|$.

Next it is shown that $|D|$ divides $\phi(F)$. For convenience, let $\bar{F}$ denote $F - K$. Then $N_r D^{-1} N_l = \bar{F}$. Since N_r, and D are right-coprime, there exist $X, Y \in \mathbf{M}(\mathbf{R})$ such that $XN_r + YD = I$. Similarly, since D and N_l are left-coprime, there exist $\tilde{X}, \tilde{Y} \in \mathbf{M}(\mathbf{R})$ such that $N_l\tilde{X} + D\tilde{Y} = I$. Now

$$D^{-1} = D^{-1}(N_l\tilde{X} + D\tilde{Y}) = D^{-1}N_l\tilde{X} + \tilde{Y}, \tag{4.3.7}$$
$$N_r D^{-1} = N_r D^{-1}N_l\tilde{X} + N_r\tilde{Y} = \bar{F}\tilde{X} + N_r\tilde{Y}. \tag{4.3.8}$$

Similarly,

$$D^{-1} = (XN_r + YD)D^{-1}$$
$$= XN_r D^{-1} + Y$$
$$= X\bar{F}\tilde{X} + XN_r\tilde{Y} + Y. \tag{4.3.9}$$
$$|D| = \phi(D^{-1}) \text{ by Lemma 4.3.8}$$
$$= \phi(X\bar{F}\tilde{X} + XN_r\tilde{Y} + Y) \text{ by } (4.3.9)$$
$$= \phi(X\bar{F}\tilde{X}) \text{ by Lemma 4.3.5}, \tag{4.3.10}$$

since $XN_r\tilde{Y} + Y \in \mathbf{M}(\mathbf{R})$. However, by Lemma 4.3.6, $\phi(X\bar{F}\tilde{X})$ divides $\phi(\bar{F}) = \phi(F - K) = \bar{F}$, using Lemma 4.3.5 again. Hence, $|D|$ divides $\phi(F)$. □

Corollary 4.3.9 *Suppose $F \in \mathbf{M}(\mathbf{F})$, and that (N_r, D, N_l, K) is a bicoprime factorization of F. Then $F \in \mathbf{M}(\mathbf{R})$ if and only if $D \in \mathbf{U}(\mathbf{R})$.*

Corollary 4.3.10   *Suppose $F \in \mathbf{M}(\mathbf{F})$, and let (N, D), $(\tilde{D}, \tilde{N})$ be any r.c.f. and any l.c.f. of F. Then $|D|$ and $|\tilde{D}|$ are both characteristic determinants of F.*

Proof. Note that $(N, D, I, 0)$ and $(I, \tilde{D}, \tilde{N}, 0)$ are both bicoprime factorizations of F, and apply Theorem 4.3.4. □

The rest of this section is devoted to extending Fact 2.1.3 to rational matrices, by introducing the notion of McMillan degree. Suppose $F \in \mathbf{M}(\mathbb{R}(s))$ and that $p \in C$ is a pole of $F(\cdot)$; the *McMillan degree* of p as a pole of F is the highest order it has as a pole of any minor of F.

Example 4.3.11   Consider again the rational matrix F of (4.3.2). This matrix has three poles, namely 0, 1, and -1. The McMillan degrees of 0 and -1 are 1, while the McMillan degree of 1 is 2.

The following important result is now obtained almost routinely from the preceding discussion.

Theorem 4.3.12   Suppose $F \in \mathbf{M}(\mathbb{R}(s))$, and suppose (N, D), $(\tilde{D}, \tilde{N})$, (C, A, B, K) are any r.c.f., any l.c.f. and any bicoprime factorization of F over the ring $\mathbf{S}$. Then
   (i) $|D|, |\tilde{D}|, |A|$ are associates.
   (ii) An element $p \in C_{+e}$ is a pole of F if and only if it is a zero of a characteristic determinant of F, and its McMillan degree as a pole of F equals its multiplicity as a zero of $\phi(F)$.
   (iii) An element $p \in C_{+e}$ is a pole of F if and only if $|D(p)| = 0$ (or equivalently $|\tilde{D}(p)| = 0$ or $|A(p)| = 0$). If $p \in C_{+e}$ is a pole of F, its McMillan degree equals its multiplicity as a zero of $|D|$ (or $|\tilde{D}|$ or $|A|$).

4.4   MATRIX EUCLIDEAN DIVISION

Throughout this section, $\mathbf{R}$ denotes a *proper* Euclidean domain. Recall that a domain $\mathbf{R}$ is *Euclidean* if there is a "degree" function $\delta : \mathbf{R} \setminus 0 \to \mathbf{Z}_+$ such that

   (i) Given any $f, g \in \mathbf{R}$ with $g \neq 0$, there exists a $q \in \mathbf{R}$ such that either $r := f - gq$ is zero or else $\delta(r) < \delta(g)$.

   (ii) If x divides y, then $\delta(x) \leq \delta(y)$.

A Euclidean domain $\mathbf{R}$ is *proper* if it is not a field, and $\delta(fg) = \delta(f) + \delta(g)$ whenever f, g are nonzero. However, even in a proper Euclidean domain, the "quotient" q and "remainder" r corresponding to a given pair f, g need not be unique: this is so in the ring $\mathbf{S}$, for example. Accordingly, given $f, g \in \mathbf{R} \setminus 0$, define

$$I(f, g) = \min_{q \in \mathbf{R}} \delta(f - gq), \tag{4.4.1}$$

where $\delta(0)$ is taken as $-\infty$. Clearly $I(f, g) < \delta(g)$, and it is either nonnegative or $-\infty$. In the case $\mathbf{R} = \mathbf{S}$, Theorem 2.3.2 provides an expression for $I(\cdot, \cdot)$. The objective of this section is to answer the following question: Given $A, B \in \mathbf{M}(\mathbf{R})$ with A square and A, B right-coprime, over what elements of $\mathbf{R}$ does $|A + RB|$ vary? This question arises in the study of the strong stabilizability of multivariable plants. It turns out that the answer to this question is quite simple, though its derivation is rather involved.

Theorem 4.4.1   Suppose $A, B \in \mathbf{M}(\mathbf{R})$ are right-coprime and A is square. Let $a = |A|$ and let b denote the smallest invariant factor of B. Then the sets

$$\{a + rb : r \in \mathbf{R}\}, \tag{4.4.2}$$

$$\{|A + RB| : R \in \mathbf{M}(\mathbf{R})\}, \tag{4.4.3}$$

are equal. As a consequence,

$$\min_{R \in \mathbf{M}(\mathbf{R})} \delta(|A + RB|) = \min_{r \in \mathbf{R}} \delta(a + rb) = I(a, b). \tag{4.4.4}$$

Remarks 4.4.2   The first part of the theorem means that, if any element $f \in \mathbf{R}$ can be expressed as $a + rb$ for some $r \in \mathbf{R}$, then there exists an $R \in \mathbf{M}(\mathbf{R})$ such that $f = |A + RB|$, and conversely. The relation (4.4.4) follows readily once the equality of the sets in (4.4.2) and (4.4.3) is established.

Proof.   The proof is divided into two parts: First, it is shown that, corresponding to every $R \in \mathbf{M}(\mathbf{R})$, there exists an $r \in \mathbf{R}$ such that $|A + RB| = a + rb$. Then it is shown that, corresponding to every $r \in \mathbf{R}$, there exists an $R \in \mathbf{M}(\mathbf{R})$ such that $a + rb = |A + RB|$.

To prove the first assertion, suppose $R \in \mathbf{M}(\mathbf{R})$ is selected arbitrarily. Note that

$$A + R = [I \ R] \begin{bmatrix} A \\ B \end{bmatrix}. \tag{4.4.5}$$

Using (4.4.5) and the Binet-Cauchy formula (Fact B.1.5), one can obtain an expansion for $|A + RB|$ as a sum involving products of minors of $[I \ R]$ and of $F = [A' \ B']'$. Now, except for $|A|$, every other minor of F contains at least one row from B and is hence a multiple of b. In the Binet-Cauchy expansion, $|A|$ is multiplied by $|I|$, which is one. Hence, $|A + RB| = a + rb$ for an appropriate $r \in \mathbf{R}$.

To prove the second assertion, suppose $r \in \mathbf{R}$ is selected arbitrarily; it is then necessary to find an $R \in \mathbf{M}(\mathbf{R})$ such that $a + rb = |A + RB|$. This is first proved under the assumption that $|A| \neq 0$; this assumption is removed later on. Let A^{adj} denote the adjoint matrix of A, and let $c_1, \cdots, c_k$ denote the invariant factors of BA^{adj} (where $k = \mathrm{rank}\, BA^{adj} = \mathrm{rank}\, BA^{-1} = \mathrm{rank}\, B$). The major part of the proof consists of demonstrating the existence of an $R_0 \in \mathbf{M}(\mathbf{R})$ such that $|aI + R_0 C| = (a + rb)a^{n-1}$, where $n \times n$ is the size of A, and

$$
C = \begin{bmatrix} c_1 & & & 0 \\ & \cdot & & \\ & & \cdot & \\ & & & \cdot & \\ 0 & & & c_k \\ & & 0 & & 0 \end{bmatrix},
\tag{4.4.6}
$$

is a Smith form of BA^{adj}. If this can be shown, then the existence of an $R \in \mathbf{M}(\mathbf{R})$ such that $|A + RB| = a + rb$ follows readily: Select $U, V \in \mathbf{U}(\mathbf{R})$ such that $UBA^{adj}V = C$, and define

$$
R = V R_0 U .
\tag{4.4.7}
$$

To show that $|A + RB| = a + rb$, it is enough to show that $|A + RB| \cdot |A^{adj}| = (a + rb)a^{n-1}$, since $|A^{adj}| = a^{n-1}$. But

$$
\begin{aligned}
|A + RB| \cdot |A^{adj}| &= |V^{-1}(A + RB)A^{adj}V| \\
&= |aI + R_0 UBA^{adj}V| \\
&= |aI + R_0 C| \\
&= a^{n-1}(a + rb) ,
\end{aligned}
\tag{4.4.8}
$$

provided R_0 satisfies $|aI + R_0 C| = (a + rb)a^{n-1}$.

To demonstrate the existence of such R_0, let

$$
\begin{bmatrix} s_1/t_1 & & & 0 \\ & \cdot & & \\ & & \cdot & \\ & & & \cdot \\ & & & s_k/t_k \\ 0 & & & 0 \end{bmatrix}
\tag{4.4.9}
$$

be a Smith-McMillan form of BA^{-1}. Since B, A are right-coprime, (B, A) is an r.c.f. of BA^{-1}. Hence, from the proof of Theorem 4.1.15, it follows that

$$
B \sim \begin{bmatrix} S & 0 \\ 0 & 0 \end{bmatrix}, \quad A \sim \begin{bmatrix} T & 0 \\ 0 & I_{n-k} \end{bmatrix},
\tag{4.4.10}
$$

where $S = \mathrm{Diag}\,\{s_1, \cdots, s_k\}$, $T = \mathrm{Diag}\,\{t_1, \cdots, t_k\}$. In particular, we have

$$
b \sim s_1, \quad a \sim \prod_{i=1}^{k} t_i .
\tag{4.4.11}
$$

Without loss of generality, it is assumed that the equivalences in (4.4.11) are actually equalities. Now BA^{adj} is equivalent to C defined in (4.4.6), so that Ca^{-1} is a Smith-McMillan form for BA^{-1}. Hence,

$$\frac{s_i}{t_i} = \frac{c_i}{a} \text{ for } i = 1, \cdots, k , \tag{4.4.12}$$

where again equivalence has been replaced by equality. Thus,

$$c_1 = \frac{s_1}{t_1} \cdot a = s_1 t_2 \cdots t_k ,$$
$$c_1 c_2 = \frac{s_1 s_2}{t_1 t_2} \cdot a^2 = (s_1 s_2 t_3 \cdots t_k)a$$
$$\vdots$$
$$c_1 c_2 \cdots c_k = \frac{s_1 s_2 \cdots s_k}{t_1 t_2 \cdots t_k} a^k = (s_1 \cdots s_k)a^{k-1} . \tag{4.4.13}$$

Since s_i, t_i are coprime for all i, it is straight-forward to show that

$$\text{g.c.d. } \{t_2 t_3 \cdots t_k, s_2 t_3 \cdots t_k, s_2 s_3 \cdots t_k, \cdots, s_2 s_3 \cdots s_k\} = 1 . \tag{4.4.14}$$

Hence, from (4.4.13) and (4.4.14),

$$\text{g.c.d. } \{a^{k-1}c_1, a^{k-2}c_1 c_2, \cdots, c_1 c_2 \cdots c_k\} = a^{k-1}s_1 = a^{k-1}b . \tag{4.4.15}$$

In view of (4.4.15), there exist elements $q_1, \cdots, q_k \in \mathbf{R}$ such that

$$q_1 a^{k-1}c_1 + q_2 a^{k-2}c_1 c_2 + \cdots + q_k c_1 c_2 \cdots c_k = a^{k-1}br . \tag{4.4.16}$$

Now define R_0 as the bordered companion matrix

$$R_0 = \begin{bmatrix} q_1 & q_2 & \cdots & q_k & 0 & \cdots & 0 \\ -1 & 0 & & & \vdots & & \\ 0 & -1 & & & 0 & & \\ & & & & & & \\ & & & & 0 & & \end{bmatrix} . \tag{4.4.17}$$

It is claimed that $|aI + R_0 C| = (a + rb)a^{n-1}$. To see this, note that $aI + R_0 C$ is the sum of a diagonal matrix and a nondiagonal matrix. Using Fact B.1.6 it is possible to expand $|aI + R_0 C|$ as a sum of products of principal minors of aI and of $R_0 C$. This gives

$$|aI + R_0 C| = a^n + \sum_{i=1}^{k} a^{n-i} \left(\sum i \times i \text{ principal minors of } R_0 C \right) , \tag{4.4.18}$$

because every $i \times i$ minor of aI equals a^i, and all minors of $R_0 C$ larger than $k \times k$ equal zero (recall that C has rank k). Now, since R_0 is a companion matrix, the only nonzero $i \times i$ principal minor of R_0 is the one corresponding to rows (and columns) 1 to i; it equals q_i. All other $i \times i$ principal minors of R_0 equal zero. Hence, there is only one nonzero $i \times i$ principal minor of $R_0 C$, and it equals $q_i c_1 \cdots c_i$. Finally, from (4.4.18) and (4.4.16),

$$|aI + R_0 C| = a^n + a^{n-k}\left(a^{k-1}q_1 c_1 + a^{k-2}q_2 c_1 c_2 + \cdots + q_k c_1 \cdots c_k\right)$$

$$= a^n + a^{n-k} \cdot a^{k-1} br = a^{n-1}(a + rb) . \qquad (4.4.19)$$

It only remains to address the case where $|A| = 0$ in order to complete the proof. In this case, since A, B are right-coprime, it follows from Problem 4.1.3 that b is a unit; so let $b = 1$, without loss of generality. Then $\{a + rb : r \in \mathbf{R}\} = \mathbf{R}$. The conclusion of the proof requires a result of independent interest, so it is displayed separately.

Lemma 4.4.3 *Suppose $P, Q \in \mathbf{M}(\mathbf{R})$ have the same number of columns, and that $T = [P'\ Q']'$ has full column rank. Then there exists an $R \in \mathbf{M}(\mathbf{R})$ such that $|P + RQ| \neq 0$.*

Proof. If $|P| \neq 0$ let $R = 0$. Otherwise select a nonzero full-size minor of T having as few rows from Q as possible. Such a minor must contain at least one row of Q since $|P| = 0$. Let t denote this minor, and suppose it is obtained by excluding rows $i_1, \cdots, i_k$ of P and including rows $j_1, \cdots, j_k$ of Q. By assumption, *every* minor of T containing fewer than k rows of Q equals zero. Now define $R \in \mathbf{M}(\mathbf{R})$ by

$$r_{i_1 j_1} = \cdots = r_{i_k j_k} = 1 ; \quad r_{ij} = 0 \text{ for all other } i, j . \qquad (4.4.20)$$

Then every $k \times k$ minor of R equals zero, except for the minor consisting of rows $i_1, \cdots, , i_k$ and columns $j_1, \cdots, j_k$, which equals one. Every larger minor of R equals zero. Next, observe that

$$P + RQ = [I \ R]\begin{bmatrix} P \\ Q \end{bmatrix} . \qquad (4.4.21)$$

Hence, using the Binet-Cauchy formula (Fact B.1.5), one can expand $|P + RQ|$ in terms of the minors of $[I \ R]$ and of T. Further, by Laplace's expansion (Fact B.1.4), every minor of $[I \ R]$ equals $\pm$ a corresponding minor of R. In particular, every minor of $[I \ R]$ containing more than k columns of R equals zero. There is only one nonzero minor of $[I \ R]$ containing exactly k columns of R, and it is obtained by omitting columns $i_1, \cdots, i_k$ of the identity matrix and including columns $j_1, \cdots, j_k$ of R; it equals ± 1. Since every minor of T containing fewer than k rows of Q equals zero, we finally get

$$|P + RQ| = \pm t , \qquad (4.4.22)$$

which is nonzero. $\square$

Proof of Theorem 4.4.1 (conclusion). Suppose $|A| = 0$. Then $b = 1$, as shown previously. Suppose $r \in \mathbf{R}$ is specified. First select $R_0 \in \mathbf{M}(\mathbf{R})$ such that $|A + R_0 B| \neq 0$; such an R_0 exists, by Lemma 4.4.3. Define $\bar{A} = A + R_0 B$, $\bar{a} = |\bar{A}|$. By previous discussion, there exists an $R_1 \in \mathbf{M}(\mathbf{R})$ such that $|\bar{A} + R_1 B| = \bar{a} + (r - \bar{a})b = r$. Now let $R = R_0 + R_1$. $\square$

Corollary 4.4.4   *Suppose $A, B \in \mathbf{M}(\mathbf{R})$ are right-coprime and A is square. Let b denote the smallest invariant factor of B. If b is not a unit, then $|A + RB| \neq 0$ for all $R \in \mathbf{M}(\mathbf{R})$; moreover, $|A + RB|$ and b are coprime for all $R \in \mathbf{M}(\mathbf{R})$.*

Proof. Let $a = |A|$. In view of Theorem 4.4.1, it is enough to show that $a + rb \neq 0$ for all $r \in \mathbf{R}$ and that $a + rb, b$ are coprime for all $r \in \mathbf{R}$. By Problem 2.1.3, a and b are coprime, since A and B are right-coprime. The desired conclusions now follow readily.   □

Corollary 4.4.5   *Suppose $A, B \in \mathbf{M}(\mathbf{R})$ are both square, with $|B| \neq 0$. Then there exists an $R \in \mathbf{M}(\mathbf{R})$ such that*

$$\delta(|A + RB|) < \delta(|B|) . \tag{4.4.23}$$

Proof. If $|A| = 0$, (4.4.23) is satisfied with $R = 0$, so suppose $|A| \neq 0$. Let F be a g.c.r.d. of A, B, and let $A = A_1 F, B = B_1 F$. Let $a_1 = |A_1|$ and let b_1 denote the smallest invariant factor of B_1. Then Theorem 4.4.1 implies that, for some $R \in \mathbf{M}(\mathbf{R})$,

$$\begin{aligned}
\delta(|A + RB|) &= \delta(|F|) + \delta(|A_1 + RB_1|) \\
&= \delta(|F|) + I(a_1, b_1) \\
&< \delta(|F|) + \delta(b_1) \\
&\leq \delta(|F|) + \delta(|B_1|) = \delta(|B|) .
\end{aligned} \tag{4.4.24}$$

This completes the proof.   □

NOTES AND REFERENCES

Most of the material in Section 4.1 is standard. Kailath [53] introduces many of these results for matrices with elements in the polynomial ring $\mathbb{R}[s]$ or the field $\mathbb{R}(s)$, but his methods largely carry over to the general situation studied here. See also [106]. The factorization in Section 4.2 is from [72], which extends earlier results from [96]. The results concerning matrix Euclidean division in Section 4.4 are taken from [102].

CHAPTER 5

Stabilization

This is one of the central chapters in the book, whose theme is the stabilization of one or more plants using one or more compensators. The first step is a study of the problem of stabilizing a (not necessarily stable) plant using an appropriate compensator, and the main result is a parametrization of *all* compensators that stabilize a given plant. This parametrization enables one to answer questions such as the following: (i) When can a plant be stabilized using a stable compensator? (ii) When can a single compensator stabilize each of two or more plants? (iii) When can a plant be stabilized using a multicompensator configuration? (iv) What are the advantages of a "two degrees of freedom" compensator? (v) Once a plant has been stabilized, what type of performance can be expected from it in tasks such as tracking, disturbance rejection, etc.?

Throughout the chapter, the symbol **S**, introduced in Chapter 2, is used to denote the set of proper stable rational functions. The symbol **M(S)** (resp. **U(S)**) denotes the set of matrices (resp. unimodular matrices) whose elements belong to **S**.

5.1 CLOSED-LOOP STABILITY

Consider the feedback system shown in Figure 5.1, where P represents the plant and C the compensator; u_1, u_2 denote the externally applied inputs, e_1, e_2 denote the inputs to the compensator and plant respectively, and y_1, y_2 denote the outputs of the compensator and plant respectively. This model is versatile enough to accomodate several control problems. For instance, in a problem of tracking, u_1 would be a reference signal to be tracked by the plant output y_2. In a problem of disturbance rejection or desensitization to noise, u_1 would be the disturbance/noise. Depending on whether u_1 *or* u_2 is the externally applied control signal (as opposed to noise etc.), the configuration of Figure 5.1 can represent either feedback or cascade compensation. For convenience, we will refer to this set-up as a feedback system.

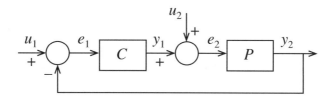

Figure 5.1: Feedback System.

Suppose $P, C \in \mathbf{M}(\mathbb{R}(s))$ so that they represent lumped linear time-invariant (continuous-time) systems, and are of compatible dimensions so that the interconnection in Figure 5.1 makes sense. The system under study is then described by

$$\begin{bmatrix} e_1 \\ e_2 \end{bmatrix} = \begin{bmatrix} u_1 \\ u_2 \end{bmatrix} - \begin{bmatrix} 0 & P \\ -C & 0 \end{bmatrix} \begin{bmatrix} e_1 \\ e_2 \end{bmatrix} ; \quad \begin{bmatrix} y_1 \\ y_2 \end{bmatrix} = \begin{bmatrix} C & 0 \\ 0 & P \end{bmatrix} \begin{bmatrix} e_1 \\ e_2 \end{bmatrix} . \tag{5.1.1}$$

This system of equations can be rewritten as

$$e = u - FGe ; \quad y = Ge , \tag{5.1.2}$$

where

$$e = \begin{bmatrix} e_1 \\ e_2 \end{bmatrix}, \quad u = \begin{bmatrix} u_1 \\ u_2 \end{bmatrix}, \quad y = \begin{bmatrix} y_1 \\ y_2 \end{bmatrix}, \quad F = \begin{bmatrix} 0 & I \\ -I & 0 \end{bmatrix}, \quad G = \begin{bmatrix} C & 0 \\ 0 & P \end{bmatrix} . \tag{5.1.3}$$

It is easy to verify that $|I + FG| = |I + PC| = |I + CP|$. The system (5.1.1) is *well-posed* if this determinant is nonzero as an element of $\mathbb{R}(s)$, i.e., if $|I + (FG)(s)|$ is not identically zero for all s. This condition is necessary and sufficient to ensure that (5.1.1) that has a unique solution in $\mathbf{M}(\mathbb{R}(s))$ for e_1, e_2 corresponding to every $u_1, u_2 \in \mathbf{M}(\mathbb{R}(s))$ of appropriate dimension.

If the system (5.1.1) is well-posed, then (5.1.1) can be solved for e_1, e_2; this gives

$$e = (I + FG)^{-1} u =: H(P, C)u . \tag{5.1.4}$$

It is possible to obtain several equivalent expressions for $H(P, C)$. Perhaps the most easily proved is

$$H(P, C) = \begin{bmatrix} (I + PC)^{-1} & -P(I + CP)^{-1} \\ C(I + PC)^{-1} & (I + CP)^{-1} \end{bmatrix} . \tag{5.1.5}$$

In (5.1.5) both $(I + PC)^{-1}$ and $(I + CP)^{-1}$ occur, which is not very convenient. To get around this, one can use the following well-known matrix identities [26]:

$$(I + PC)^{-1} = I - P(I + CP)^{-1}C, \quad C(I + PC)^{-1} = (I + CP)^{-1}C . \tag{5.1.6}$$

Of course (5.1.6) also holds with P and C interchanged throughout. Using these identities gives two other expressions for $H(P, C)$, namely

$$H(P, C) = \begin{bmatrix} I - P(I + CP)^{-1}C & -P(I + CP)^{-1} \\ (I + CP)^{-1}C & (I + CP)^{-1} \end{bmatrix}$$

$$= \begin{bmatrix} (I + PC)^{-1} & -(I + PC)^{-1}P \\ C(I + PC)^{-1} & I - C(I + PC)^{-1}P \end{bmatrix} . \tag{5.1.7}$$

Of these, the first involves only $(I + CP)^{-1}$ and the second only $(I + PC)^{-1}$.

The pair (P, C) is *stable* if $|I + PC| = |I + CP| \neq 0$ and $H(P, C) \in \mathbf{M(S)}$. Under these conditions, we also say that the feedback system (5.1.1) is stable. Put into other words, the system (5.1.1) is stable if it is well-posed, and if the transfer matrix from u to e belongs to $\mathbf{M(S)}$. The reader can verify that the conditions for stability are symmetric in P and C; thus (P, C) is stable if and only if (C, P) is stable.

A natural question at this stage is: Why should stability be defined in terms of the transfer matrix from u to e? Why not in terms of the transfer matrix from u to y? The answer is that both notions of stability are equivalent.

Lemma 5.1.1 *Suppose the system* (5.1.1) *is well-posed, and define* $W(P, C) = G(I + FG)^{-1}$, *where* G *and* F *are as in* (5.1.3). *Then* $W(P, C) \in \mathbf{M(S)}$ *if and only if* $H(P, C) \in \mathbf{M(S)}$.

Proof. Observe that F is nonsingular and that $F^{-1} = -F$. Hence,

$$H = I - FW, \quad W = F^{-1}(I - H) = F(H - I), \tag{5.1.8}$$

where H, W are shorthand for $H(P, C), W(P, C)$ respectively. It is immediate from (5.1.8) that $W \in \mathbf{M(S)}$ if and only if $H \in \mathbf{M(S)}$. $\qquad\square$

In some undergraduate textbooks the feedback system (5.1.1) is defined to be "stable" if $P(I + CP)^{-1} \in \mathbf{M(S)}$. Thus, one may ask why the current definition of stability requires *four different* transfer matrices to belong to $\mathbf{M(S)}$. There are two answers: (i) If $C \in \mathbf{M(S)}$, then $H(P, C) \in \mathbf{M(S)}$ if and only if $P(I + CP)^{-1} \in \mathbf{M(S)}$. Thus, if one uses a stable compensator, then the stability of (5.1.1) can be ascertained by examining only $P(I + CP)^{-1}$. (ii) In the general case where no assumptions are made about C, "$H(P, C) \in \mathbf{M(S)}$" is a necessary and sufficient condition for "internal" as well as "external" stability. Each of these points is elaborated below.

Lemma 5.1.2 *Suppose* $C \in \mathbf{M(S)}$. *Then* $H(P, C) \in \mathbf{M(S)}$ *if and only if* $P(I + CP)^{-1} \in \mathbf{M(S)}$.

Proof. "only if" Obvious.

"if" Suppose $P(I + CP)^{-1} \in \mathbf{M(S)}$, and label the four matrices in (5.1.7) as H_{11} through H_{22}. Thus, the hypothesis is that $-H_{12} \in \mathbf{M(S)}$. Clearly this implies that $H_{12} \in \mathbf{M(S)}$. Next,

$$H_{11} = I - P(I + CP)^{-1}C = I + H_{12}C \in \mathbf{M(S)} \tag{5.1.9}$$

since both H_{12} and C belong to $\mathbf{M(S)}$. This in turn implies that

$$H_{21} = CH_{11} \in \mathbf{M(S)}. \tag{5.1.10}$$

Finally,

$$\begin{aligned} H_{22} &= I - C(I + PC)^{-1}P = I - CP(I + PC)^{-1} \\ &= I + CH_{12} \in \mathbf{M(S)}. \end{aligned} \tag{5.1.11}$$

Thus, $H(P, C) \in \mathbf{M(S)}$. $\qquad\square$

Suppose now that no assumptions are made regarding the stability of C. In order to discuss the internal stability of the feedback system (5.1.1), suppose C and P are described by state-space equations, of the form

$$\dot{x}_c = Q_c x_c + R_c u_c; \quad y_c = S_c x_c + T_c u_c , \tag{5.1.12}$$
$$\dot{x}_p = Q_p x_p + R_p u_p; \quad y_p = S_p x_p + T_p u_p , \tag{5.1.13}$$

Lemma 5.1.3   *Suppose the triples (S_c, Q_c, R_c), (S_p, Q_p, R_p) are both stabilizable and detectable, and that $|I + C(\infty)P(\infty)| = |I + T_c T_p| \neq 0$. Under these conditions, the system of Figure 5.1 is asymptotically stable if and only if $H(P, C) \in \mathbf{M(S)}$.*

Remarks 5.1.4   Lemma 5.1.3 shows that internal stability and external stability are equivalent if the state-space representations of C and P are stabilizable and detectable. Note that the representations of C and P need not be minimal; but any "hidden modes" must be stable.

The proof of Lemma 5.1.3 uses the following well-known result, which can be found in [53].

Lemma 5.1.5   *Suppose a triple $(\bar{S}, \bar{Q}, \bar{R})$ is stabilizable and detectable, and that T is a matrix of compatible dimensions. Under these conditions, the system*

$$\dot{x}(t) = \bar{Q}x(t) , \tag{5.1.14}$$

is asymptotically stable if and only if

$$W(s) = \bar{S}(sI - \bar{Q})^{-1}\bar{R} + \bar{T} , \tag{5.1.15}$$

belongs to $\mathbf{M(S)}$.

Proof of Lemma 5.1.3. Define block-diagonal matrices Q, R, S, T in the obvious manner, and let F be as in (5.1.3). Then it is routine to verify that the interconnected system is described by the equations

$$\dot{x} = \bar{Q}x + \bar{R}u, \quad y = \bar{S}x + \bar{T}u , \tag{5.1.16}$$

where

$$\begin{aligned}
\bar{Q} &= Q - R(I + TF)^{-1}S , \\
\bar{R} &= R[I - F(I + TF)^{-1}T] = R(I + FT)^{-1} , \\
\bar{S} &= (I + TF)^{-1}S , \\
\bar{T} &= (I + TF)^{-1}T .
\end{aligned} \tag{5.1.17}$$

It is now shown that the triple $(\bar{S}, \bar{Q}, \bar{R})$ is stabilizable and detectable. Recall that $(\bar{Q}, \bar{R})$ is stabilizable if and only if there exists a constant matrix $\bar{K}$ such that $\bar{Q} - \bar{R}\bar{K}$ is Hurwitz [78].[1] By assumption, the pair (Q, R) is stabilizable, so that there exists a matrix K such that $Q - RK$ is Hurwitz. Now define

$$\bar{K} = (I + FT)(K - \bar{S}) . \tag{5.1.18}$$

Then

$$\bar{Q} - \bar{R}\bar{K} = Q - R\bar{S} - RK + R\bar{S} = Q - RK , \tag{5.1.19}$$

which is Hurwitz by construction. This shows that the pair $(\bar{Q}, \bar{R})$ is stabilizable. The detectability is shown in the same manner.

To conclude the proof, observe that, by Lemma 5.1.5, the overall system is asymptotically stable (or "internally" stable) if and only if its transfer matrix $W(P, C) \in \mathbf{M}(\mathbf{S})$. However, by Lemma 5.1.1 the latter condition is equivalent to the requirement that $H(P, C) \in \mathbf{M}(\mathbf{S})$ (i.e., the system is "externally" stable). $\qquad\square$

Having defined the notion of stability for feedback systems and examined its implications, we now move on to the main objective of this section, which is to characterize the stability of the system (5.1.1) in terms of coprime factorizations over $\mathbf{S}$ of the plant and compensator. This is done next in Theorem 5.1.6, which is the principal result of this section.

Theorem 5.1.6 Suppose $P, C \in \mathbf{M}(\mathbb{R}(s))$. Let (N_p, D_p), $(\tilde{D}_p, \tilde{N}_p)$ be any r.c.f. and any l.c.f. of P, and let (N_c, D_c), $(\tilde{D}_c, \tilde{N}_c)$ be any r.c.f. and l.c.f. of C. Under these conditions, the following are equivalent:
(i) The pair (P, C) is stable.
(ii) The matrix $\tilde{N}_c N_p + \tilde{D}_c, D_p$ is unimodular.
(iii) The matrix $\tilde{N}_p N_c + \tilde{D}_p, D_c$ is unimodular.

The proof is based on the following lemma, which is of independent interest.

Lemma 5.1.7 *Let all symbols be as in Theorem 5.1.6, and suppose the pair (P, C) is well-posed. Define*

$$\Delta(P, C) = \tilde{N}_c N_p + \tilde{D}_c D_p , \quad \tilde{\Delta}(P, C) = \tilde{N}_p N_c + \tilde{D}_p D_c . \tag{5.1.20}$$

Then both $|\Delta(P, C)|$ and $|\tilde{\Delta}(P, C)|$ are characteristic determinants of $H(P, C)$ (in the sense of Definition 4.3.2).

---

[1] A matrix is *Hurwitz* if all of its eigenvalues have negative real parts.

Proof. It is only shown that $|\Delta(P, C)|$ is a characteristic determinant of $H(P, C)$. The parallel assertion for $|\tilde{\Delta}(P, C)|$ is proved in an entirely analogous fashion.

Note that $\Delta = \Delta(P, C) = \tilde{D}_c(I + CP)D_p$. If the pair (P, C) is well-posed, then $|I + CP| \neq 0$, and by definition $|\tilde{D}_c| \neq 0$, $|D_p| \neq 0$. Hence, $|\Delta| \neq 0$, and from (5.1.7),

$$H(P, C) = \begin{bmatrix} I - N_p\Delta^{-1}\tilde{N}_c & -N_p\Delta^{-1}\tilde{D}_c \\ D_p\Delta^{-1}\tilde{N}_c & D_p\Delta^{-1}\tilde{D}_c \end{bmatrix}$$

$$= \begin{bmatrix} I & 0 \\ 0 & 0 \end{bmatrix} + \begin{bmatrix} -N_p \\ D_p \end{bmatrix}\Delta^{-1}[\tilde{N}_c \ \tilde{D}_c]. \tag{5.1.21}$$

Let K denote the first (constant) matrix on the right side of (5.1.21). It is claimed that the quadruple $([-N'_p \ D'_p]', \Delta, [\tilde{N}_c\tilde{D}_c], K)$ is a bicoprime factorization of $H(P, C)$. To show this, it is only necessary to show that

$$\begin{bmatrix} -N_p \\ D_p \\ \Delta \end{bmatrix} \sim \begin{bmatrix} I \\ 0 \end{bmatrix}, \quad [\tilde{N}_c \ \tilde{D}_c \ \Delta] \sim [I \ 0]. \tag{5.1.22}$$

But this is immediate from the right-coprimeness of N_p, D_p and the left-coprimeness of $\tilde{N}_c$, $\tilde{D}_c$: Any common right divisor of $-N_p$, D_p, Δ must, in particular, be a common right divisor of $-N_p$, D_p and hence must be unimodular; the second part of (5.1.22) is proved in the same manner. Now by Theorem 4.3.4 we conclude that $|\Delta(P, C)|$ is a characteristic determinant of $H(P, C)$. □

Proof of Theorem 5.1.6. (i) $\implies$ (ii) Suppose the pair (P, C) is stable. Then in particular it is well-posed. By Lemma 5.1.7 and Corollary 4.3.9, it follows that $\Delta(P, C) \in \mathbf{U}(\mathbf{S})$.

(ii) $\implies$ (i) Suppose $\Delta(P, C) \in \mathbf{U}(\mathbf{S})$. Since $\Delta(P, C) = \tilde{D}_c(I + CP)D_p$, it follows that $|I + CP| \neq 0$ so that the pair (P, C) is well-posed. Further, (5.1.21) shows that $H(P, C) \in \mathbf{M}(\mathbf{S})$, since $[\Delta(P, C)]^{-1} \in \mathbf{M}(\mathbf{S})$.

The proof that (i) and (iii) are equivalent is entirely similar. □

Let us say that a compensator C *stabilizes* the plant P if the pair (P, C) is stable. Then Theorem 5.1.6 leads to the result below.

Corollary 5.1.8   *Suppose $P \in \mathbf{M}(\mathbb{R}(s))$, and let (N_p, D_p), $(\tilde{D}_p, \tilde{N}_p)$ be any r.c.f. and any l.c.f. of P. Then the following are equivalent:*
(i) C stabilizes P.
(ii) C has an l.c.f. $(\tilde{D}_c, \tilde{N}_c)$ such that $\tilde{D}_c D_p + \tilde{N}_c N_p = I$.
(iii) C has an r.c.f. (N_c, D_c) such that $\tilde{D}_p D_c + \tilde{N}_p N_c = I$.

Proof. Once again, only the equivalence of (i) and (ii) is demonstrated; the equivalence of (i) and (iii) follows along similar lines.

(i) $\implies$ (ii) Since I is unimodular, it follows from Theorem 5.1.6 that (P, C) is stable, i.e., that C stabilizes P.

(ii) $\Longrightarrow$ (i) Suppose C stabilizes P, and let $(\tilde{D}, \tilde{N})$ be any l.c.f. of C. By Theorem 5.1.6, the matrix $\Delta := \tilde{D}D_p + \tilde{N}N_p$, is unimodular. Now let $\tilde{D}_c = \Delta^{-1}\tilde{D}$, $\tilde{N}_c = \Delta^{-1}\tilde{N}$. Then by the left analog of Theorem 4.1.13, $(\tilde{D}_c, \tilde{N}_c)$ is also an l.c.f. of C, and moreover, $\tilde{D}_c D_p + \tilde{N}_c N_p = \Delta^{-1}(\tilde{D}D_p + \tilde{N}N_p) = I$.

$\square$

Remarks 5.1.9   Let P be a plant, and let (N_p, D_p), $(\tilde{D}_p, \tilde{N}_p)$ be any r.c.f. and any l.c.f. of P. Then every C that stabilizes P has a *unique* l.c.f. $(\tilde{D}_c, \tilde{N}_c)$ such that $\tilde{D}_c D_p + \tilde{N}_c N_p = I$. Moreover, in terms of this *particular* r.c.f. (N_p, D_p) of P and l.c.f. $(\tilde{D}_c, \tilde{N}_c)$ of C, we have

$$H(P, C) = \begin{bmatrix} I - N_p\tilde{N}_c & -N_p\tilde{D}_c \\ D_p\tilde{N}_c & D_p\tilde{D}_c \end{bmatrix}. \tag{5.1.23}$$

This follows readily from (5.1.7). Similarly,

$$W(P, C) = F^{-1}(I - H(P, C)) = \begin{bmatrix} D_p\tilde{N}_c & D_p\tilde{D}_c - I \\ N_p\tilde{N}_c & N_p\tilde{N}_c \end{bmatrix}. \tag{5.1.24}$$

The "right" analogs of (5.1.23) and (5.1.24) are easy to derive and are left to the reader.

PROBLEMS

5.1.1. (a) Suppose P, C are given, and let (N_p, D_p), (N_c, D_c) be any r.c.f.'s of P and C, respectively. Show that the pair (P, C) is stable if and only if

$$U = \begin{bmatrix} D_c & -N_p \\ N_c & D_p \end{bmatrix} \in \mathbf{U(S)}\ .$$

(b) State and prove the "left" analog of (a).

5.1.2. The objective of this problem is to show that in a stable feedback system there can be no unstable pole-zero cancellations between P and C.

(a) Suppose P, C are scalar, and that (n_p, d_p), (n_c, d_c) are coprime factorizations of P and C. Suppose the pair (P, C) is stable. Show that no C_{+e}-pole of P can be a zero of C and vice versa. (Hint: Show that n_p, d_c are coprime and that n_c, d_p are coprime.)

(b) Prove the following multivariable generalization of (a): If (P, C) is stable, then N_p, D_c are left-coprime and N_c, D_p are left-coprime. (Hint: See Problem 5.1.1.)

(c) State and prove the left analog of (b).

5.2   PARAMETRIZATION OF ALL STABILIZING COMPENSATORS

Let $P \in \mathbf{M}(\mathbb{R}(s))$ and let $S(P)$ denote the set of all $C \in \mathbf{M}(\mathbb{R}(s))$ that stabilize P, i.e., the set of all $C \in \mathbf{M}(\mathbb{R}(s))$ such that (P, C) is stable. The main objective of this section is to parametrize the set $S(P)$, as well as the set $\{H(P, C) : C \in S(P)\}$. Thus, we wish to parametrize the set of all compensators that stabilize a given plant P, together with the set of all stable closed-loop transfer matrices that can be generated from P by using an appropriate stabilizing compensator.

Theorem 5.2.1   Suppose $P \in \mathbf{M}(\mathbb{R}(s))$, and let (N_p, D_p), $(\tilde{D}_p, \tilde{N}_p)$ be any r.c.f. and any l.c.f. of P. Select matrices $X, Y, \tilde{X}, \tilde{Y} \in \mathbf{M}(\mathbf{S})$ such that $XN_p + YD_p = I$, $\tilde{N}_p\tilde{X} + \tilde{D}_p\tilde{Y} = I$. Then

$$
\begin{aligned}
S(P) &= \left\{ (Y - R\tilde{N}_p)^{-1}(X + R\tilde{D}_p) : R \in \mathbf{M}(\mathbf{S}) , \quad |Y - R\tilde{N}_p| \neq 0 \right\} \\
&= \left\{ (\tilde{X} + D_pS)(\tilde{Y} - N_pS)^{-1} : S \in \mathbf{M}(\mathbf{S}) , \quad |\tilde{Y} - N_pS| \neq 0 \right\} .
\end{aligned}
\tag{5.2.1}
$$

Remarks 5.2.2   Equation (5.2.1) gives two representations for the set $S(P)$ of compensators that stabilize the plant P. The first of these states the following: Let R be any element of $\mathbf{M}(\mathbf{S})$ of appropriate dimensions such that $|Y - R\tilde{N}_p| \neq 0$; then the corresponding compensator $C = (Y - R\tilde{N}_p)^{-1}(X + R\tilde{D}_p)$ stabilizes P. Conversely, suppose $C \in \mathbf{M}(\mathbb{R}(s))$ stabilizes P; then C is of the form $(Y - R\tilde{N}_p)^{-1}(X + R\tilde{D}_p)$ for some $R \in \mathbf{M}(\mathbf{S})$ (and it is implicit that $|Y - R\tilde{N}_p| \neq 0$). The second representation has a similar interpretation.

Proof.   Only the first representation is proved below. The proof of the second is entirely similar and is left to the reader.

It is shown in Corollary 5.1.8 that C stabilizes P if and only if C has an l.c.f. $(\tilde{D}_c, \tilde{N}_c)$ such that $\tilde{D}_c\tilde{D}_p + \tilde{N}_cN_p = I$. Now consider the equation

$$
\bar{X}N_p + \bar{Y}D_p = I ,
\tag{5.2.2}
$$

in the unknowns $\bar{X}$ and $\bar{Y}$. Then C stabilizes P if and only if C is of the form $\bar{Y}^{-1}\bar{X}$ for some $\bar{X}, \bar{Y} \in \mathbf{M}(\mathbf{S})$ such that (5.2.2) holds and $|\bar{Y}| \neq 0$. From Corollary 4.1.18, the matrix

$$
U_1 = \begin{bmatrix} Y & X \\ -\tilde{N}_p & \tilde{D}_p \end{bmatrix} ,
\tag{5.2.3}
$$

is unimodular, and moreover U_1^{-1} is of the form

$$
U_1^{-1} = \begin{bmatrix} D_p & \\ & G \\ N_p & \end{bmatrix} ,
\tag{5.2.4}
$$

where $G \in \mathbf{M}(\mathbf{S})$. By Corollary 4.1.18, it now follows that all solutions for $(\bar{Y}, \bar{X})$ of (5.2.2) are of the form

$$[\bar{Y} \ \bar{X}] = [I \ R]U_1 = [Y - R\tilde{N}_p \ X + R\tilde{D}_p] \text{ for some } R \in \mathbf{M}(\mathbf{S}) . \qquad (5.2.5)$$

The first representation in (5.2.1) is implied by (5.2.5). □

Theorem 5.2.1 characterizes the set of all compensators that stabilize a given plant in terms of a certain "free" parameter. The correspondence between the parameter and the compensator is one-to-one in the following sense: Suppose P is a given plant, and that we have chosen a *particular* r.c.f. (N_p, D_p), l.c.f. $(\tilde{D}_p, \tilde{N}_p)$, as well as particular matrices $X, Y, \tilde{X}, \tilde{Y} \in \mathbf{M}(\mathbf{S})$ such that $XN_p + YD_p = I, \tilde{N}_p\tilde{X} + \tilde{D}_p\tilde{Y} = I$; then, corresponding to each $C \in S(P)$, there is a *unique* $R \in \mathbf{M}(\mathbf{S})$ such that $C = (Y - R\tilde{N}_p)^{-1}(X + R\tilde{D}_p)$, as well as a *unique* $S \in \mathbf{M}(\mathbf{S})$ such that $C = (\tilde{X} + D_pS)(\tilde{Y} - N_pS)^{-1}$. This fact is easy to prove and is left to the reader as an exercise (see Problem 5.2.1).

Theorem 5.2.1 shows that there is a simple formula that generates all the stabilizing compensators for a given plant. By substituting this formula into the expression for $H(P, C)$, we can obtain a parametrization of all stable closed-loop transfer matrices that can be obtained by stabilizing a given plant. The bonus is that this parametrization is *affine* in the free parameters R or S.

Corollary 5.2.3 *Suppose $P \in \mathbf{M}(\mathbb{R}(s))$ and let $N_p, D_p, \tilde{N}_p, \tilde{D}_p, X, Y, \tilde{X}, \tilde{Y}$ be as in Theorem 5.2.1. Suppose $C \in S(P)$. Then*

$$H(P, C) = \begin{bmatrix} I - N_p(X + R\tilde{D}_p) & -\tilde{N}_p(Y - R\tilde{N}_p) \\ D_p(X + R\tilde{D}_p) & D_p(Y - R\tilde{N}_p) \end{bmatrix}, \qquad (5.2.6)$$

where R is the unique element of $\mathbf{M}(\mathbf{S})$ such that $C = (Y - R\tilde{N}_p)^{-1}(X + R\tilde{D}_p)$; alternatively,

$$H(P, C) = \begin{bmatrix} (\tilde{Y} - N_pS)\tilde{D}_p & -(\tilde{Y} - N_pS)\tilde{N}_p \\ (\tilde{X} + D_pS)\tilde{D}_p & I - (\tilde{X} + D_pS)\tilde{N}_p \end{bmatrix}, \qquad (5.2.7)$$

where S is the unique element of $\mathbf{M}(\mathbf{S})$ such that $C = (\tilde{X} + D_pS)(\tilde{Y} - N_pS)^{-1}$. Similarly,

$$\begin{aligned} W(P, C) &= \begin{bmatrix} D_p(X + R\tilde{D}_p) & D_p(Y - R\tilde{N}_p) - I \\ N_p(X + R\tilde{D}_p) & N_p(Y - R\tilde{N}_p) \end{bmatrix}, \\ &= \begin{bmatrix} (\tilde{X} + D_pS)\tilde{D}_p & (\tilde{X} + D_pS)\tilde{N}_p \\ (\tilde{Y} - N_pS)\tilde{D}_p & -I(\tilde{Y} - N_pS)\tilde{N}_p \end{bmatrix}. \end{aligned} \qquad (5.2.8)$$

The proof of the corollary is left to the reader; it is a ready consequence of Theorem 5.2.1 and Remark 5.1.9.

If $C \in S(P)$, it follows from (5.2.8) that the transfer matrices from u_1 to y_2 and from u_2 to y_2 are both right multiples of N_p, no matter which stabilizing compensator is used. This is a multivariable generalization of a fact well-known in the scalar case, namely that if an unstable plant is stabilized by feedback compensation, then the nonminimum phase zeros of the plant continue to be zeros of the resulting stable closed-loop transfer function, irrespective of which stabilizing compensator is used.

In the characterization (5.2.1) of $S(P)$, the parameters R and S are not entirely free, since the nonsingularity constraint in (5.2.1) must be respected. It is shown next that "almost all" R and S satisfy this constraint.

Lemma 5.2.4 below can be thought of as an extension of Lemma 4.4.3 to topological rings.

Lemma 5.2.4   *Suppose $A, B \in \mathbf{M(S)}$ have the same number of columns, that A is square, and that $F = [A' B']'$ has full (column) rank. Define*

$$\mathbf{V} = \{R \in \mathbf{M(S)} : |A + RB| \neq 0\} . \tag{5.2.9}$$

Then $\mathbf{V}$ is an open dense subset of $\mathbf{M(S)}$.

Proof. It is necessary to prove two things: (i) If $R \in \mathbf{V}$, then there is a neighborhood $\mathbf{N}(R)$ of R such that $\mathbf{N}(R) \subseteq \mathbf{V}$ (this shows that $\mathbf{V}$ is open). (ii) If $R \notin \mathbf{V}$, then there is a sequence $\{R_i\}$ in $\mathbf{V}$ such that $R_i \to R$ (this shows that $\mathbf{V}$ is dense).

Accordingly, suppose first that $R \in \mathbf{V}$, so that $|A + RB| =: d \neq 0$. Observe that if $\varepsilon < \|d\|$, then 0 does not belong to the ball $\mathbf{B}(d, \varepsilon)$ defined by

$$\mathbf{B}(d, \varepsilon) = \{x \in \mathbf{S} : \|x - d\| < \varepsilon\} . \tag{5.2.10}$$

Also note that the function $R \mapsto f(R) = |A + RB|$ is a continuous map from $\mathbf{M(S)}$ into $\mathbf{S}$. Thus, the preimage $f^{-1}(\mathbf{B}(d, \varepsilon))$ is an open neighborhood of R, and is a subset of $\mathbf{V}$. This shows that $\mathbf{V}$ is open.

Next, suppose that $R \notin \mathbf{V}$ so that $|A + RB| = 0$. Define $\bar{A} = A + RB$ and observe that $\bar{F} := [\bar{A}' B']'$ has full column rank since

$$\begin{bmatrix} \bar{A} \\ B \end{bmatrix} = \begin{bmatrix} I & R \\ 0 & I \end{bmatrix} \begin{bmatrix} A \\ B \end{bmatrix} . \tag{5.2.11}$$

It is shown that there exists a sequence $\{C_i\}$ in $\mathbf{M(S)}$ converging to zero such that $|\bar{A} + C_i B| \neq 0 \,\forall i$. Setting $R_i = R + C_i$ completes the proof that $\mathbf{V}$ is dense.

The rest of the proof is very reminiscent of that of Lemma 4.4.3. Select a nonzero full-size minor of $\bar{F}$ containing as few rows of B as possible. Let $\bar{f}$ denote this minor and suppose it is obtained by omitting rows $i_1, \cdots, i_k$ of A and including rows $j_1, \cdots, j_k$ of B. Define $C^\varepsilon \in \mathbf{M(S)}$ by

$$c^\varepsilon_{i_1 j_1} = \cdots = c^\varepsilon_{i_k j_k} = \varepsilon, \quad c^\varepsilon_{ij} = 0 \text{ for all other } i, j . \tag{5.2.12}$$

Then, as in the proof of Lemma 4.4.3,

$$|\bar{A} + C^\varepsilon B| = \pm \varepsilon^k \bar{f} \text{ if } \varepsilon \neq 0 . \tag{5.2.13}$$

Now let $\{\varepsilon_i\}$, be any sequence of nonzero numbers converging to zero, and let $C_i = C^{\varepsilon_i}$. Then $C_i \to 0$ but $|\bar{A} + C_i B| \neq 0$ for all i.   □

Corollary 5.2.5   *With all symbols as in Theorem 5.2.1, the set of $R \in \mathbf{M}(\mathbf{S})$ such that $|Y - R\tilde{N}_p| \neq 0$ is an open dense subset of $\mathbf{M}(\mathbf{S})$, and the set of $S \in \mathbf{M}(\mathbf{S})$ such that $|\tilde{Y} - N_p S| \neq 0$ is an open dense subset of $\mathbf{M}(\mathbf{S})$.*

Proof. Since the matrix U_1 of (5.2.3) is unimodular, it follows in particular that Y and $\tilde{N}_p$ are right-coprime. Hence, the matrix $F = [Y' \ \tilde{N}_p']'$ has full column rank. Now apply Lemma 5.2.4.
□

Lemma 5.2.6   *Let all symbols be as in Theorem 5.2.1, and let n_p denote the smallest invariant factor of $\tilde{N}_p$. Suppose n_p is not a unit. Then*

$$S(P) = \{(Y - R\tilde{N}_p)^{-1}(X + R\tilde{D}_p) : R \in \mathbf{M}(\mathbf{S})\}$$
$$= \{(\tilde{X} + D_p S)(\tilde{Y} - N_p S)^{-1} : S \in \mathbf{M}(\mathbf{S})\} . \tag{5.2.14}$$

Moreover, every $C \in S(P)$ is analytic at every C_{+e}-zero of n_p.

Proof. By Corollary 4.4.4, if n_p is not a unit, then $|Y - R\tilde{N}_p| \neq 0$ for *all* $R \in \mathbf{M}(\mathbf{S})$, and $|Y - R\tilde{N}_p|, n_p$ are coprime.   □

Lemma 5.2.6 can be interpreted as follows: An $s \in C_{+e}$ where $P(s) = 0$ is called a *blocking zero* of P. It is easily established that the C_{+e}-blocking zeros of P are precisely the C_{+e}-zeros of n_p (see Problem 5.2.3). The lemma states that if P has any C_{+e}-blocking zeros, then the "nonsingularity constraint" in (5.2.1) is automatically satisfied and can therefore be omitted (compare (5.2.1) and (5.2.14)). Moreover, every C that stabilizes P is analytic at the C_{+e}-blocking zeros of P.

If P is strictly proper, then infinity is a blocking zero of P. In this case Lemma 5.2.6 leads to the following result:

Corollary 5.2.7   *Suppose P is strictly proper. Then every $C \in S(P)$ is proper.*

Lemma 5.2.6 assumes a particularly simple form if P is stable.

Corollary 5.2.8   *Let all symbols be as in Lemma 5.2.6. Suppose $P \in \mathbf{M}(\mathbf{S})$ and that n_p is not a unit. Then*

$$S(P) = \{(I - RP)^{-1}R : R \in \mathbf{M}(\mathbf{S})\}$$
$$= \{R(I - PR)^{-1} : R \in \mathbf{M}(\mathbf{S})\} . \tag{5.2.15}$$

In this case

$$H(P, C) = \begin{bmatrix} I - PR & -P(I - RP) \\ R & I - RP \end{bmatrix}, \tag{5.2.16}$$

$$W(P, C) = \begin{bmatrix} R & -RP \\ PR & P(I - RP) \end{bmatrix}. \tag{5.2.17}$$

Proof. Since P is stable, select $N_p = \tilde{N}_p = P$, $D_p = I$, $\tilde{D}_p = I$, $Y = I$, $\tilde{Y} = I$, $X = \tilde{X} = 0$, and apply Lemma 5.2.6 together with the formula (5.2.6). This establishes (5.2.16). Similarly (5.2.17) follows from Lemma 5.2.6 and (5.2.8). $\qquad\square$

In particular, if P is stable and has some C_{+e}-blocking zeros, then the set of transfer matrices from u_1 to y_2 that can be obtained by applying a stabilizing compensator is precisely the set of right multiples of P.

The next lemma shows that, given any plant P and any *finite* set of points in C_{+e} that is disjoint from the set of poles of P, there exists a stabilizing compensator for P that has a blocking zero at each of these points. The main motivation for this lemma is to show that every proper plant can be stabilized by a strictly proper compensator. However, the general result follows with very little more effort.

Lemma 5.2.9   *Given a plant P and points $s_1, \cdots, s_n \in C_{+e}$ such that none is a pole of P, there exists a $C \in S(P)$ such that $C(s_i) = 0$ for all i.*

Proof. Select an $f \in \mathbf{S}$ such that $f(s_i) = 0 \, \forall i$. Let (N_p, D_p), $(\tilde{D}_p, \tilde{N}_p)$ be any r.c.f. and any l.c.f. of P, and select any $X, Y \in \mathbf{M}(\mathbf{S})$ such that $XN_p + YD_p = I$. Since none of the s_i is a pole of P, none is a zero of $|\tilde{D}_p|$ (see Theorem 4.3.12 (iii)). Hence, f can be selected so that $|\tilde{D}_p|$, f are coprime. Let d denote $|\tilde{D}_p|$ for convenience. Since d, f are coprime, for every $x \in \mathbf{S}$ there exist $a, b, \in \mathbf{S}$ such that $ad + bf = x$; equivalently, for every $x \in \mathbf{S}$ there exists an $a \in \mathbf{S}$ such that $x + ad$ is divisible by f. Accordingly, select $A \in \mathbf{M}(\mathbf{S})$ such that every element of $X + Ad$ is divisible by f, and let $R = A\tilde{D}_p^{adj}$. Then $X + R\tilde{D}_p = X + Ad$ is a multiple of f and vanishes at each s_i. Moreover,

$$(Y - R\tilde{N}_p)D_p + (X + R\tilde{D}_p)N_p = I. \tag{5.2.18}$$

Finally, $|Y - R\tilde{N}_p| \neq 0$ since $[(Y - R\tilde{N}_p)D_p](s_i) = I$ for all i. Thus, $(Y - R\tilde{N}_p)^{-1}$ is well-defined, and $C = (Y - R\tilde{N}_p)^{-1}(X + R\tilde{D}_p)$ is in $S(P)$ and satisfies $C(s_i) = 0$ for all i. $\qquad\square$

Proposition 5.2.10 below rounds out the contents of this section; it characterizes *all* compensators that satisfy the conditions of Lemma 5.2.9.

Proposition 5.2.10   *Given a plant P and points $s_1, \cdots, s_n \in C_{+e}$ such that none of the s_i is a pole of P, let (N_p, D_p), $(\tilde{D}_p, \tilde{N}_p)$ be any r.c.f. and l.c.f. of P. Let $\mathbf{I}$ denote the ideal in $\mathbf{S}$ consisting of all functions* *that vanish at each of the s_i, and let f denote a generator of this ideal. In accordance with Lemma 5.2.9, select $X, Y, \tilde{X}, \tilde{Y} \in \mathbf{M}(\mathbf{S})$ such that*

$$XN + YD = I, \quad \tilde{N}\tilde{X} + \tilde{D}\tilde{Y} = I, \tag{5.2.19}$$
$$X, \tilde{X} \in \mathbf{M}(\mathbf{I}), \tag{5.2.20}$$

where the last equation merely states that both X and $\tilde{X}$ vanish at each s_i. Then the set of all $C \in S(P)$ that vanish at each s_i is given by

$$\{(Y - R\tilde{N}_p)^{-1}(X + R\tilde{D}_p) : R \in \mathbf{M}(\mathbf{I})\}$$
$$= \{(\tilde{X} + D_p S)(\tilde{Y} - N_p S)^{-1} : S \in \mathbf{M}(\mathbf{I})\}. \tag{5.2.21}$$

Proof. Suppose $C = (Y - R\tilde{N}_p)^{-1}(X + R\tilde{D}_p)$ where $R \in \mathbf{M}(\mathbf{I})$. Then $X + R\tilde{D}_p \in \mathbf{M}(\mathbf{I})$, so that $C(s_i) = 0 \, \forall i$.

Conversely, suppose $C \in S(P)$ and $C(s_i) = 0 \, \forall i$. Then, by Theorem 5.2.1, C equals $(Y - R\tilde{N}_p)^{-1}(X + R\tilde{D}_p)$ for some $R \in \mathbf{M}(\mathbf{S})$, and all that remains to be shown is that this R actually belongs to $\mathbf{M}(\mathbf{I})$. Now, since C vanishes at each s_i, it follows that $X + R\tilde{D}_p \in \mathbf{M}(\mathbf{I})$. Since $X \in \mathbf{M}(\mathbf{I})$, this implies successively that $R\tilde{D}_p \in \mathbf{M}(\mathbf{I})$, and that $R|\tilde{D}_p| = R\tilde{D}_p\tilde{D}_p^{adj} \in \mathbf{M}(\mathbf{I})$. Since $|\tilde{D}_p|$ coprime to f, this implies that $R \in \mathbf{M}(\mathbf{I})$. $\qquad\square$

As a final observation, suppose it is desired to determine all compensators C such that $H(P, C) \in \mathbf{M}(\mathbf{S_D})$, where $\mathbf{D}$ is a prespecified subset of the complex plane. Then all of the results of this section carry over completely, provided $\mathbf{S}$ is replaced by $\mathbf{S_D}$, and C_{+e} is replaced by the extended complement of $\mathbf{D}$, i.e., the union of $\{\infty\}$ and the complement of $\mathbf{D}$. An illustration of this is provided in Section 3.1. In the same manner, the results of this section can be extended in a completely transparent way to discrete-time systems.

PROBLEMS

5.2.1. In the parametrization (5.2.1) of $S(P)$, show that distinct choices of R lead to distinct compensators. (Hint: Suppose $(Y - R_1\tilde{N}_p)^{-1}(X + R_1\tilde{D}_p) = (Y - R_2\tilde{N}_p)^{-1}(X + R_2\tilde{D}_p)$ for some $R_1, R_2 \in \mathbf{M}(\mathbf{S})$. Show first the existence of a $U \in \mathbf{U}(\mathbf{S})$ such that

$$[Y - R_1\tilde{N}_p \ \ X + R_1\tilde{D}_p] = U[Y - R_2\tilde{N}_p \ \ X + R_2\tilde{D}_p].$$

Write this in the form

$$[I \ \ R_1] = \begin{bmatrix} Y & X \\ -\tilde{N}_p & \tilde{D}_p \end{bmatrix} = U[I \ \ R_2] \begin{bmatrix} Y & X \\ -\tilde{N}_p & \tilde{D}_p \end{bmatrix}.$$

Now use Corollary 4.1.18 to show that $U = I$, $R_1 = R_2$.)

5.2.2. In (5.2.1), suppose the matrices $X, Y, \tilde{X}, \tilde{Y}$ also satisfy $X\tilde{Y} = Y\tilde{X}$ (such matrices can always be found, by Theorem 4.1.16). Show that $C_1 = (Y - R\tilde{N}_p)^{-1}(X + R\tilde{D}p)$ equals $C_2 = (\tilde{X} + D_p S)(\tilde{Y} - N_p S)^{-1}$ if and only if $R = S$.

5.2.3. Suppose $P \in \mathbf{M}(\mathbb{R}(s))$ and let (N_p, D_p), $(\tilde{D}_p, \tilde{N}_p)$ be any r.c.f. and l.c.f. of P. Let n_p denote the smallest invariant factor of N_p (and of $\tilde{N}_p$; see Theorem 4.1.15). Show that the C_{+e}-blocking zeros of P are precisely the C_{+e}-zeros of n_p.

5.3   STRONG STABILIZATION

The question studied in this section is the following: Given an unstable plant P, when does there exist a *stable* compensator C that stabilizes P? More generally, what is the minimum number of C_{+e}-poles (counted according to their McMillan degrees) that any stabilizing compensator for P must have? A plant P is *strongly stabilizable* if $S(P)$ contains a stable compensator. A study of strong stabilizability is of interest for its own sake; some consequences of strong stabilizability are discussed towards the end of this section. In addition, several other problems in reliable stabilization are intimately related to strong stabilizability (see Sections 5.4 and 5.5). It turns out that there exist very simple necessary and sufficient conditions for strong stabilizability, based on the numbers and locations of the real C_{+e}-poles and zeros of the plant.

Theorem 5.3.1   Given $P \in \mathbf{M}(\mathbb{R}(s))$, let $\sigma_1, \cdots, \sigma_l$ denote the real C_{+e}-blocking zeros of P (including ∞ if P is strictly proper), arranged in ascending order. Let η_i denote the number of poles of P in the interval (σ_i, σ_{i+1}), counted according to their McMillan degrees, and let η denote the number of odd integers in the sequence $\{\eta_1, \cdots, \eta_{l-1}\}$. Then every $C \in S(P)$ has at least η poles in C_{+e}. Moreover, this lower bound is exact in that there is a $C \in S(P)$ with exactly η poles in C_{+e}.

Corollary 5.3.2   *P is strongly stabilizable if and only if the number of poles of P (counted according to their McMillan degrees) between any pair of real C_{+e}-blocking zeros of P is even.*

Remarks 5.3.3   Both Theorem 5.3.1 and Corollary 5.3.2 are natural generalizations of the corresponding scalar results (Theorem 3.2.1 and Corollary 3.2.2, respectively). As in the scalar case, the condition for strong stabilizability given in Corollary 5.3.2 is referred to as the *parity interlacing property (p.i.p.)*.

Proof of Theorem 5.3.1.   By Theorem 5.2.1, every $C \in S(P)$ is of the form $(Y - R\tilde{N}_p)^{-1}(X + R\tilde{D}_p)$ for some $R \in \mathbf{M}(\mathbf{S})$ such that $|Y - R\tilde{N}_p| \neq 0$. By Theorem 4.2.1, the sum of the McMillan degrees of the C_{+e}-poles of C equals $\delta(Y - R\tilde{N}_p)$. Thus, the theorem is proved if it can be shown that

$$\min_{R \in \mathbf{M}(\mathbf{S}), \, |Y - R\tilde{N}_p| \neq 0} \delta(Y - R\tilde{N}_p) = \eta . \tag{5.3.1}$$

Let n_p denote the smallest invariant factor of $\tilde{N}_p$. There are two cases to consider, namely where n_p is a unit and where it is not. If n_p is a unit then $\eta = 0$. Also, by Theorem 4.4.1, the quantity $|Y - R\tilde{N}_p|$ can made to equal every element of $\mathbf{S}$ by an appropriate choice of R. In particular, there exists an $R \in \mathbf{M}(\mathbf{S})$ such that $\delta(Y - R\tilde{N}_p) = 0$. Since $\delta(f) \geq 0 \; \forall f \in \mathbf{S}$, (5.3.1) holds in the case where n_p is a unit.

If n_p is not a unit, then by Corollary 4.4.4, $|Y - R\tilde{N}_p| \neq 0$ for all $R \in \mathbf{M}(\mathbf{S})$. Hence, the nonsingularity constraint in (5.3.1) can be dropped, and (5.3.1) simplifies to

$$\min_{R \in \mathbf{M}(\mathbf{S})} \delta(|Y - R\tilde{N}_p|) = \eta \; . \tag{5.3.2}$$

Now by Theorem 4.4.1,

$$\min_{R \in \mathbf{M}(\mathbf{S})} \delta(|Y - R\tilde{N}_p|) = \min_{r \in \mathbf{M}(\mathbf{S})} \delta(y - rn_p) \; , \tag{5.3.3}$$

where $y = |Y|$. Now note that the C_{+e}-zeros of n_p are precisely the C_{+e}-blocking zeros of P (see Problem 5.2.3). Hence, by Theorem 2.3.2, the minimum on the right side of (5.3.3) equals the number of sign changes in the sequence $\{y(\sigma_1), \cdots, y(\sigma_l)\}$. Next, observe that, since $XN_p + YD_p = I$ and $N_p(\sigma_i) = 0$, we have $(YD_p)(\sigma_i) = I$ for all i. This shows that $y(\sigma_i)$ and $|D_p(\sigma_i)|$ have the same sign for all i. Thus, the minimum in (5.3.3) equals the number of sign changes in the sequence $\{|D_p(\sigma_1)|, \cdots, |D_p(\sigma_l)|\}$. Finally, note that $|D_p(\sigma_i)|$ and $|D_p(\sigma_{i+1})|$ have opposite signs if and only if the number of zeros of $|D_p(\cdot)|$ in the interval (σ_i, σ_{i+1}) is odd. By Theorem 4.2.1, the zeros of $|D_p(\cdot)|$ are in one-to-one correspondence with the poles of P, and the multiplicity of a zero of $|D_p(\cdot)|$ is the same as its McMillan degree as a pole of P. This completes the proof. $\qquad \square$

The proof of the corollary is straight-forward and is left to the reader.

Note that the condition of Corollary 5.3.2 is vacuously satisfied if P has at most one real C_{+e}-blocking zero, of if P has no real C_{+e}-poles.

The following alternate form of Corollary 5.3.2 is sometimes useful.

Corollary 5.3.4   *Suppose $P \in \mathbf{M}(\mathbb{R}(s))$, and let (N_p, D_p) be any r.c.f. of P. Then P is strongly stabilizable if and only if $|D_p(\cdot)|$ has the same sign at all real C_{+e}-blocking zeros of P.*

The next result gives a canonical form for a strongly stabilizable plant. This form is not very useful as a test for strong stabilizability (such a test is provided by Corollary 5.3.2), but is very useful for proving theorems about strongly stabilizable systems.

Lemma 5.3.5   *A plant $P \in \mathbf{M}(\mathbb{R}(s))$ is strongly stabilizable if and only if there exist $C, V \in \mathbf{M}(\mathbf{S})$ such that $(V, I - CV), (I - VC, V)$ are respectively an r.c.f. and an l.c.f. of P.*

Proof. "if" Suppose $(V, I - CV)$ is an r.c.f. of P. Then, since

$$C \cdot V + I \cdot (I - CV) = I \,, \tag{5.3.4}$$

it follows that C stabilizes P. Hence, P is strongly stabilizable.

"only if" Suppose P is strongly stabilizable. Select $C \in \mathbf{M(S)}$ such that $P(I + CP)^{-1} =: V \in \mathbf{M(S)}$. It is now easy to verify that $P = (I - VC)^{-1}V = V(I - CV)^{-1}$, and that $(I - VC, V)(V, I - CV)$ are respectively left-coprime and right-coprime. $\square$

It is now possible to explain one of the essential differences between a strongly stabilizable plant and one that is not. Refer to (5.1.24), which gives an expression for all possible $W(P, C)$ that can be generated by a stabilizing compensator C. If P is strongly stabilizable and if $C \in S(P)$ is chosen to be stable, then $\tilde{D}_c$ is unimodular. As a result, N_p and $W_{22}(P, C) = N_p \tilde{D}_c$ have the same invariant factors. To put it another way, if P is stabilized using a *stable* compensator, then the resulting stable transfer matrix has the same C_{+e}-zeros as the original plant P. On the other hand, if P is stabilized using an *unstable* compensator (and this is all one can do if P is not strongly stabilizable), then $|\tilde{D}_c|$ is not a unit, so that at least one invariant factor of $N_p \tilde{D}_c$ is a strict multiple of the corresponding invariant factor of N_p. In other words, stabilization using an unstable compensator always introduces additional C_{+e}-zeros into the closed-loop transfer matrix beyond those of the original plant. As we will see in Chapter 6, the RHP zeros of a plant affect its ability to track reference signals and/or to reject disturbances. Hence, it is preferable to use a stable stabilizing compensator in such situations.

In some design problems, one uses a so-called *two-stage procedure* for selecting an appropriate compensator. Given a plant P, the first stage consists of selecting a stabilizing compensator for P. Let $C \in S(P)$ denote this compensator and define $P_1 = P(I + CP)^{-1}$. The second stage consists of selecting a stabilizing compensator for P_1 that also achieves some other design objectives such as decoupling, sensitivity minimization, etc. The rationale behind this procedure is that the design problems are often easier to solve when the plant is stable. The resulting configuration with its inner and outer loops is shown in Figure 5.2. (Note that the system of Figure 5.1 has been slightly rearranged.)

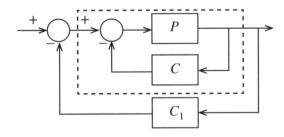

Figure 5.2: Two-Stage Compensator Design.

However, in order for this two-stage procedure to be valid, it must be true that

$$S(P) = C + S(P_1) , \tag{5.3.5}$$

or, in other words, C_1 stabilizes P_1 if and only if $C + C_1$ stabilizes P. This is not always true in general.

Theorem 5.3.6   Suppose $P \in \mathbf{M}(\mathbb{R}(s))$, $C \in S(P)$, and let P_1 denote $P(I + CP)^{-1}$. Then

$$C + S(P_1) \subseteq S(P) , \tag{5.3.6}$$

with equality holding if and only if $C \in \mathbf{M}(\mathbf{S})$.

Remarks 5.3.7   Theorem 5.3.6 states that, in general, if $C \in S(P)$ and $C_1 \in S(P_1)$, then $C + C_1 \in S(P)$. However, unless $C \in \mathbf{M}(\mathbf{S})$ (which is possible only if P is strongly stabilizable), there exists a C_1 such that $C + C_1 \in S(P)$ but $C_1 \notin S(P_1)$. To put it another way, the two-stage technique always yields a compensator that stabilizes the original plant P. However, not all compensators in $S(P)$ can be realized in this fashion; a necessary and sufficient condition for such a decomposition to be possible is that the inner-loop compensator C is stable (which implies, *a fortiori*, that P is strongly stabilizable).

Proof.   In the first part of the proof it is shown that if $C_1 \in S(P_1)$ then $C + C_1 \in S(P)$. By assumption, $H(P, C) \in \mathbf{M}(\mathbf{S})$ and $H(P_1, C_1) \in \mathbf{M}(\mathbf{S})$. Now

$$I + P(C + C_1) = (I + PC + PC_1) = (I + PC)[I + (I + PC)^{-1}PC_1]$$
$$= (I + PC)(I + P_1 C_1) . \tag{5.3.7}$$

Thus,

$$[I + P(C + C_1)]^{-1} = (I + P_1 C_1)^{-1}(I + PC)^{-1} \in \mathbf{M}(\mathbf{S}) . \tag{5.3.8}$$

Similarly,

$$[I + P(C + C_1)]^{-1}P = (I + P_1 C_1)^{-1}(I + PC)^{-1}P \in \mathbf{M}(\mathbf{S}) . \tag{5.3.9}$$
$$[I + (C + C_1)P]^{-1} = (I + CP)^{-1}(I + C_1 P_1) \in \mathbf{M}(\mathbf{S}) . \tag{5.3.10}$$

In order to complete the proof that $H(P_1, C_1) \in \mathbf{M}(\mathbf{S})$, it only remains to show that

$$[I + (C + C_1)P]^{-1}(C + C_1) \in \mathbf{M}(\mathbf{S}) . \tag{5.3.11}$$

From (5.3.10),

$$[I + (C + C_1)P]^{-1}C_1 = (I + CP)^{-1}(I + C_1 P_1)^{-1}C_1 \in \mathbf{M}(\mathbf{S}) , \tag{5.3.12}$$

so it only remains to show that

$$[I + (C + C_1)P]^{-1}C \in \mathbf{M}(\mathbf{S}) .$$ (5.3.13)

Now, from (5.1.24), $P_1 = N_p \tilde{D}_c$ for an appropriate r.c.f. (N_p, D_p) of P and l.c.f. $(\tilde{D}_c, \tilde{N}_c)$ of C. Hence, $P_1 C = N_p \tilde{N}_c \in \mathbf{M}(\mathbf{S})$. Further, from (5.2.16), $(I + C_1 P_1)^{-1}$ equals $I - R P_1$ for some $R \in \mathbf{M}(\mathbf{S})$. Therefore

$$\begin{aligned}
[I + (C + C_1)P]^{-1}C &= (I + CP)^{-1}(I + C_1 P_1)^{-1}C \\
&= (I + CP)^{-1}(I - R P_1)C \\
&= (I + CP)^{-1}C - (I + CP)^{-1}R P_1 C \\
&\in \mathbf{M}(\mathbf{S}) ,
\end{aligned}$$ (5.3.14)

since $(I + CP)^{-1}C, (I + CP)^{-1}, R, P_1 C \in \mathbf{M}(\mathbf{S})$. This completes the proof of (5.3.6).

In the second part of the proof, it is shown that if P is strongly stabilizable and $C \in \mathbf{M}(\mathbf{S})$ stabilizes P, then (5.3.6) holds with equality. From Lemma 5.3.5, $(I - P_1 C, P_1), (P_1 I - C P_1)$ are respectively an r.c.f. and an l.c.f. of P. Now apply Theorem 5.2.1 with

$$N_p = \tilde{N}_p = P_1, \quad D_p = I - C P_1, \quad \tilde{D}_p = I - P_1 C, \quad X = C, \quad Y = I .$$ (5.3.15)

This gives

$$S(P) = \left\{(I - R P_1)^{-1}[C + R(I - P_1 C)] : R \in \mathbf{M}(\mathbf{S}) \text{ and } |I - R P_1| \neq 0\right\} .$$ (5.3.16)

On the other hand, since $P_1 \in \mathbf{M}(\mathbf{S})$, a slight modification of Corollary 5.2.8 gives

$$S(P_1) = \{(I - R P_1)^{-1}R : R \in \mathbf{M}(\mathbf{S}) \text{ and } |I - R P_1| \neq 0\} .$$ (5.3.17)

Finally, note that

$$\begin{aligned}
(I - R P_1)^{-1}R + C &= (I - R P_1)^{-1}[R + (I - R P_1)C] \\
&= (I - R P_1)^{-1}[C + R(I - P_1 C)] .
\end{aligned}$$ (5.3.18)

This last equation shows that

$$S(P) = C + S(P_1) .$$ (5.3.19)

The last part of the proof consists of showing that if $C \notin \mathbf{M}(\mathbf{S})$, then the inclusion in (5.3.6) is strict. This is shown by constructing a $\bar{C} \in S(P)$ such that $\bar{C} - C \notin S(P_1)$. Since $C \in S(P)$, by Corollary 5.1.8 there exist an r.c.f. (N_p, D_p) of P and an l.c.f. $(\tilde{D}_c, \tilde{N}_c)$ of C such that $\tilde{D}_c D_p + \tilde{N}_c N_p = I$. Moreover, $\tilde{D}_c$ it *not* unimodular since C is unstable. Now let $\bar{C} = (1 + \alpha)C$, where α is a positive number less than $1/\|\tilde{N}_c N_p\|$. Then $(\tilde{D}_c, (1 + \alpha)\tilde{N}_c)$ is an l.c.f. of $\bar{C}$, and $(\tilde{D}_c, \alpha\tilde{N}_c)$ is an l.c.f. of $\bar{C} - C = \alpha C$. Now $\bar{C} \in S(P)$ since

$$\tilde{D}_c D_p + (1 + \alpha)\tilde{N}_c N_p = I + \alpha\tilde{N}_c N_p ,$$ (5.3.20)

is unimodular by virtue of the condition $\alpha < 1/\|\tilde{N}_c N_p\|$. On the other hand, it is claimed that $\bar{C} - C = \alpha C$ does *not* stabilize P. By (5.1.24), $P_1 = N_p \tilde{D}_c$, so that $(N_p \tilde{D}_c, I)$ is an r.c.f. of P. Since $(\tilde{D}_c, \alpha \tilde{N}_c)$ is an l.c.f. of αC, applying the stability criterion of Theorem 5.1.6 gives the "return difference" matrix

$$\Delta(P_1, \alpha C) = \tilde{D}_c + \alpha \tilde{N}_c N_p \tilde{D}_c = (I + \alpha \tilde{N}_c N_p) \tilde{D}_c . \tag{5.3.21}$$

However, this matrix Δ is not unimodular since it is a multiple of the nonunimodular matrix $\tilde{D}_c$. Hence, $\alpha C \notin S(P_1)$. $\qquad\square$

5.4   SIMULTANEOUS STABILIZATION

In this section, we consider the problem of designing a compensator that stabilizes each of a given family of plants. Specifically, suppose $P_0, \cdots, P_l$ are given plants; we would like to know whether or not there exists a common stabilizing controller for this set of plants. This is referred to as the *simultaneous stabilization problem*. It turns out that the problem of stabilizing l plants with a common compensator is equivalent to that of stabilizing $l - 1$ plants using a common *stable* compensator. In particular, the problem of simultaneously stabilizing two plants is equivalent to that of stabilizing a single plant using a stable compensator, which is studied in the previous section.

The simultaneous stabilization problem arises naturally in the synthesis of control systems with "integrity." In this set-up, P_0 corresponds to the nominal description of the plant to be controlled, while $P_1, \cdots, P_l$ correspond to the plant description after some sort of structural change, such as a loss of sensors, actuators, transducers, etc. The problem of synthesizing a compensator that continues to stabilize the plant even after a contingency of this sort is precisely that of finding a compensator that stabilizes each of the plants $P_0, \cdots, P_l$. Another application for this problem arises when one tries to use a compensator for a nonlinear system. As the operating point for the nonlinear system changes, so does its linearized model around its operating point. If each of these linearized models can be stabilized using a common compensator, then a fixed (i.e., independent of operating point) compensator can be used for the nonlinear system, with a saving in complexity.

To bring out the basic ideas clearly, we begin with the problem of simultaneously stabilizing two plants, which can be stated as follows: Given two plants P_0 and P_1, when does there exist a compensator C such that (P_0, C) and (P_1, C) are both stable? If such a C exists, we say that P_0, P_1 are *simultaneously stabilizable*. To facilitate the presentation of the results, we recall a few definitions.

Suppose $P \in \mathbf{M}(\mathbb{R}(s))$. Then a doubly coprime factorization of P is an r.c.f. (N, D) and an l.c.f. $(\tilde{D}, \tilde{N})$ of P, together with matrices $X, Y, \tilde{X}, \tilde{Y} \in \mathbf{M}(\mathbf{S})$ such that

$$\begin{bmatrix} Y & X \\ -\tilde{N} & \tilde{D} \end{bmatrix} \begin{bmatrix} D & -\tilde{X} \\ N & \tilde{Y} \end{bmatrix} = I . \tag{5.4.1}$$

Such matrices can always be found (see Theorem 4.1.16). Throughout the rest of this section, it is assumed that doubly coprime factorizations of P_0 and P_1 are available, and symbols such as Y_0, N_1 etc. are used without further definition.

If $C \in \mathbf{M}(\mathbf{S})$, then (C, I), (I, C) are respectively an r.c.f. and an l.c.f. of C. Hence, for a plant $P \in \mathbf{M}(\mathbb{R}(s))$, the following are equivalent (see Theorem 5.1.6):

(i) There exists $C \in \mathbf{M}(\mathbf{S})$ such that (P, C) is stable.

(ii) There exists $R \in \mathbf{M}(\mathbf{S})$ such that $D + RN$ is unimodular, where (N, D) is any r.c.f. of P.

(iii) There exists $S \in \mathbf{M}(\mathbf{S})$ such that $\tilde{D} + \tilde{N}S$ is unimodular, where $(\tilde{D}, \tilde{N})$ is any l.c.f. of P.

The main result of this section is presented next.

Theorem 5.4.1   Given two plants P_0 and P_1, define

$$A = Y_0 D_1 + X_0 N_1, \quad B = -\tilde{N}_0 D_1 + \tilde{D}_0 N_1 . \tag{5.4.2}$$

Then P_0 and P_1 can be simultaneously stabilized if and only if there exists an $M \in \mathbf{M}(\mathbf{S})$ such that $A + MB$ is unimodular.

Remarks 5.4.2   It can be shown (see Problem 5.4.1) that A and B are right-coprime. Hence, the theorem states that P_0 and P_1 can be simultaneously stabilized if and only if the associated system BA^{-1} is strongly stabilizable. Once A and B have been computed, one can use Theorem 5.4.1 to determine whether or not there exists an $M \in \mathbf{M}(\mathbf{S})$ such that $A + MB \in \mathbf{U}(\mathbf{S})$. This leads to the parity condition.

Proof. As a first step, assume that P_0 and P_1 are both strictly proper. This assumption is removed later on. The sets of compensators that stabilize P_0 and P_1 are given by (see Lemma 5.2.6)

$$S(P_0) = \{(Y_0 - R_0 \tilde{N}_0)^{-1}(X_0 + R_0 \tilde{D}_0) : R_0 \in \mathbf{M}(\mathbf{S})\} , \tag{5.4.3}$$
$$S(P_1) = \{(Y_1 - R_1 \tilde{N}_1)^{-1}(X_1 + R_1 \tilde{D}_1) : R_1 \in \mathbf{M}(\mathbf{S})\} . \tag{5.4.4}$$

Hence, P_0 and P_1 can be simultaneously stabilized if and only if there exist R_0 and R_1 such that

$$(Y_0 - R_0 \tilde{N}_0)^{-1}(X_0 + R_0 \tilde{D}_0) = (Y_1 - R_1 \tilde{N}_1)^{-1}(X_1 + R_1 \tilde{D}_1) . \tag{5.4.5}$$

Let C denote the compensator defined in the above equation. Since both sides of (5.4.5) give l.c.f.'s of C, (5.4.5) holds if and only if there is a unimodular matrix U such that

$$Y_0 - R_0 \tilde{N}_0 = U(Y_1 - R_1 \tilde{N}_1), \quad X_0 + R_0 \tilde{D}_0 = U(X_1 + R_1 \tilde{D}_1) . \tag{5.4.6}$$

Hence, P_0 and P_1 can be simultaneously stabilized if and only if there exist $R_0, R_1 \in \mathbf{M}(\mathbf{S})$, $U \in \mathbf{U}(\mathbf{S})$ such that (5.4.6) holds. Now (5.4.6) can be rewritten as

$$[I \ R_0] \begin{bmatrix} Y_0 & X_0 \\ -\tilde{N}_0 & \tilde{D}_0 \end{bmatrix} = U[I \ R_1] \begin{bmatrix} Y_1 & X_1 \\ -\tilde{N}_1 & \tilde{D}_1 \end{bmatrix} , \tag{5.4.7}$$

which is equivalent to

$$[I \ \ R_0]\begin{bmatrix} Y_0 & X_0 \\ -\tilde{N}_0 & \tilde{D}_0 \end{bmatrix}\begin{bmatrix} D_1 & -\tilde{X}_1 \\ N_1 & \tilde{Y}_1 \end{bmatrix} = U[I \ \ R_1] \,, \tag{5.4.8}$$

or

$$[I \ \ R_0]\begin{bmatrix} A & S \\ B & T \end{bmatrix} = U[I \ \ R_1] \,, \tag{5.4.9}$$

where

$$\begin{bmatrix} A & S \\ B & T \end{bmatrix} = \begin{bmatrix} Y_0 & X_0 \\ -\tilde{N}_0 & \tilde{D}_0 \end{bmatrix}\begin{bmatrix} D_1 & -\tilde{X}_1 \\ N_1 & \tilde{Y}_1 \end{bmatrix} \,. \tag{5.4.10}$$

Thus, P_0 and P_1 can be simultaneously stabilized if and only if there exist $R_0, R_1 \in \mathbf{M(S)}, U \in \mathbf{U(S)}$ such that (5.4.9) holds. Therefore the theorem is proved if we can establish that (5.4.9) holds if and only if $A + MB$ is unimodular for some $M \in \mathbf{M(S)}$.

Accordingly, suppose first that (5.4.9) holds for suitable R_0, R_1, U. Then $A + R_0B = U$ is unimodular. Conversely, suppose $A + MB$ is unimodular for some M. Then (5.4.9) holds with $U = A + MB, R_0 = M, R_1 = U^{-1}(S + R_0T)$.

To complete the proof, consider the case where P_0 and P_1 are not necessarily strictly proper. In this case, P_0 and P_1 are simultaneously stabilizable if and only if there exist $R_0, R_1 \in \mathbf{M(S)}$ such that $|Y_0 - R_0\tilde{N}_0| \neq 0, |Y_1 - R_1\tilde{N}_1| \neq 0$, and (5.4.5) holds. The preceding discussion shows that *if* (5.4.5) holds for some R_0, R_1, then $A + R_0B$ is unimodular. Hence, if P_0, P_1 can be simultaneously stabilized, then there exists an M such that $A + MB$ is unimodular. To prove the converse, suppose $A + M_0B \in \mathbf{U(S)}$ for some $M_0 \in \mathbf{M(S)}$. If we define $R_0 = M_0, R_1 = (A + M_0B)^{-1}(S + M_0T)$, then it is conceivable that $|Y_0 - R_0\tilde{N}_0|$ or $|Y_1 - R_1\tilde{N}_1|$, is zero. If so, the idea is to perturb these elements slightly without destroying the unimodularity of $A + MB$. Proceed as follows: Since the map $M \mapsto A + MB$ is continuous and since $\mathbf{U(S)}$ is an open subset of $\mathbf{M(S)}$, it follows that there is a ball $\mathbf{B}(M_0, \varepsilon)$ centered at M_0 such that $A + MB \in \mathbf{U(S)} \ \forall M \in \mathbf{B}(M_0, \varepsilon)$. Now by Lemma 5.2.4, the sets $\{R_0 : |Y_0 - R_0\tilde{N}_0| \neq 0\}, \{R_1 : |Y_1 - R_1\tilde{N}_1| \neq 0\}$ are both open dense subsets of $\mathbf{M(S)}$. Hence, for some $M \in \mathbf{B}(M_0, \varepsilon), |Y_0 - M\tilde{N}_0|$ must be nonzero, and $|Y_1 - R_1\tilde{N}_1|$ must also be nonzero with $R_1 = (A + MB)^{-1}(S + MT)$.   □

The next result gives a more intuitive interpretation of simultaneous stabilization.

Corollary 5.4.3   *Suppose P_0 is stable and P_1 is arbitrary. Then P_0 and P_1 can be simultaneously stabilized if and only if $P_1 - P_0$ is strongly stabilizable.*

Proof. It is left to the reader to give a proof based on Theorem 5.4.1 (see Problem 5.4.2); an independent proof is given below.

"if" Suppose (N_1, D_1) is an r.c.f. of P_1. Then $(N_1 - P_0D_1, D_1)$ is an r.c.f. of $P_1 - P_0$, since

$$X_1(N_1 - P_0D_1) + (Y_1 + X_1P_0)D_1 = I \,. \tag{5.4.11}$$

Suppose $R \in \mathbf{M}(\mathbf{S})$ stabilizes $P_1 - P_0$. Then from Theorem 5.1.6 it follows that

$$D_1 + R(N_1 - P_0 D_1) \in \mathbf{U}(\mathbf{S}) . \tag{5.4.12}$$

Now rearrange (5.4.12) as

$$(I - R P_0) D_1 + R N_1 \in \mathbf{U}(\mathbf{S}) , \tag{5.4.13}$$

and define $C = (I - R P_0)^{-1} R$. Then (5.4.13) shows that C stabilizes P_1. By Corollary 5.2.8, C also stabilizes P_0. Hence, P_1 and P_0 can be simultaneously stabilized.

"only if" This part consists essentially of reversing the above reasoning. Suppose P_0 and P_1 are simultaneously stabilizable, and let C be a compensator that stabilizes both. By Corollary 5.2.8, since C stabilizes P_0, C has an l.c.f. of the form $(I - R P_0, R)$ for some $R \in \mathbf{M}(\mathbf{S})$. Since C also stabilizes P_1, Theorem 5.1.6 implies (5.4.13), which in turn is the same as (5.4.12). Finally, (5.4.12) shows that R stabilizes $P_1 - P_0$, so that $P_1 - P_0$ is strongly stabilizable. □

Theorem 5.4.1 shows that the problem of simultaneously stabilizing two plants can be reduced to that of stabilizing a single plant using a stable compensator. The converse is also true: Observe that $C \in \mathbf{M}(\mathbb{R}(s))$ is stable if and only if it stabilizes the plant $P = 0$ (see (5.1.6)). Hence, the problem of stabilizing a plant P using a stable compensator is equivalent to simultaneously stabilizing 0 and P using the same compensator.

The discussion of the simultaneous stabilization of two plants is concluded with a derivation of alternate conditions that are equivalent to the one in Theorem 5.4.1.

Lemma 5.4.4   *Suppose P_0, P_1 have no common C_{+e}-poles. Then the C_{+e}-blocking zeros of $B = -\tilde{N}_0 D_1 + \tilde{D}_0 N_1$ are precisely the C_{+e}-blocking zeros of $P_1 - P_0$.*

Proof. It is first shown that if $B(s) = 0$ for some $s \in C_{+e}$, then $|D_1(s)| \neq 0$ and $|\tilde{D}_0(s)| \neq 0$. Accordingly suppose $s \in C_{+e}$ and that $B(s) = 0$. By the hypothesis, $|D_1|$ and $|\tilde{D}_0|$ are coprime, which implies that either $|D_1(s)| \neq 0$ or $|\tilde{D}_0(s)| \neq 0$. Suppose to be specific that $|D_1(s)| \neq 0$; the other case is entirely similar. Then

$$B(s) = 0 \iff (\tilde{N}_0 D_1)(s) = (\tilde{D}_0 N_1)(s)$$
$$\iff \tilde{N}_0(s) = \tilde{D}_0(s) P_1(s) . \tag{5.4.14}$$

Hence,

$$[\tilde{N}_0(s) \; \tilde{D}_0(s)] = \tilde{D}_0(s)[P_1(s) \; I] . \tag{5.4.15}$$

Now, if $|\tilde{D}_0(s)| = 0$, then $[\tilde{N}_0(s)\tilde{D}_0(s)]$ does not have full row rank. But this contradicts the left-coprimeness of the pair $(\tilde{D}_0, \tilde{N}_0)$. Hence, $|\tilde{D}_0(s)| \neq 0$. It can be similarly established that if $|\tilde{D}_0(s)| \neq 0$ then $|D_1(s)| \neq 0$.

Next, suppose $s \in C_{+e}$, and that $B(s) = 0$. It has already been established that $|\tilde{D}_0(s)| \neq 0$ and $|D_1(s)| \neq 0$. Hence, from (5.4.14), we get

$$B(s) = 0 \iff P_0(s) = P_1(s) = 0 \iff P_1(s) - P_0(s) = 0 . \tag{5.4.16}$$

This concludes the proof.   □

Lemma 5.4.5   *Suppose P_0, P_1 have no common C_{+e}-poles. Then P_0, P_1 are simultaneously stabilizable if and only if the following interlacing property holds: Let $\sigma_1, \cdots, \sigma_l$ denote the real C_{+e}-blocking zeros of the plant $P_1 - P_0$; then the number of poles of P_0 and of P_1 in every interval (σ_i, σ_{i+1}), counted according to their McMillan degrees, is even.*

Proof. By Theorem 5.4.1, P_0 and P_1 can be simultaneously stabilized if and only if the auxiliary plant BA^{-1} is strongly stabilizable. By Corollary 5.3.4, this is the case if and only if $|A|$ has the same sign at all real C_{+e}-blocking zeros of B. Next, at such a blocking zero,

$$
\begin{aligned}
A(s) &= [Y_0 D_1 + X_0 N_1](s) \\
&= [Y_0 + X_0 N_1 D_1^{-1}](s) D_1(s) \\
&= [Y_0 + X_0 N_0 D_0^{-1}](s) D_1(s) \\
&= [D_0(s)]^{-1} D_1(s) ,
\end{aligned}
\tag{5.4.17}
$$

where the last two equations follow respectively from the facts that: (i) $P_0(s) = P_1(s)$, and (ii) $Y_0 + X_0 N_0 D_0^{-1} = [Y_0 D_0 + X_0 N_0] D_0^{-1} = D_0^{-1}$. Thus, if $\sigma_1, \cdots, \sigma_l$ are the real C_{+e}-blocking zeros of $P_1 - P_0$, then P_0, P_1 can be simultaneously stabilized if and only if $|D_0(\sigma_1)|^{-1}|D_1(\sigma_1)|, \cdots, |D_0(\sigma_l)|^{-1}|D_1(\sigma_l)|$ are all of the same sign. Since $|D_0(\sigma)|^2 \geq 0 \,\forall \sigma$, this condition holds if and only if $|D_0(\sigma_1)| \cdot |D_1(\sigma_1)|, \cdots, |D_0(\sigma_l)| \cdot |D_1(\sigma_l)|$ all have the same sign. But this is true if and only if the function $|D_0(\cdot)| \cdot |D_1(\cdot)|$ has an even number of zeros in the interval (σ_i, σ_{i+1}) for all i. Since there is a one-to-one correspondence between the zeros of $|D_i(\cdot)|$ and the poles of P_i (see Theorem 4.3.12), the result follows.   □

Now we turn to the simultaneous stabilization problem with more than two plants. Given plants $P_0, \cdots, P_l$, we would like to know whether or not there exists a compensator C that stabilizes all of them. This problem can be reduced to one of simultaneously stabilizing l plants using a *stable* compensator.

Theorem 5.4.6   Suppose $P_0, \cdots, P_l$ are given plants. Define

$$
A_i = Y_0 D_i + X_0 N_i, \quad B_i = -\tilde{N}_0 D_i + \tilde{D}_0 N_i, \quad i = 1, \cdots, l .
\tag{5.4.18}
$$

Then $P_0, \cdots, P_l$ can be simultaneously stabilized if and only if there exists an $M \in \mathbf{M}(\mathbf{S})$ such that $A_i + M B_i$ is unimodular for $i = 1, \cdots, l$.

Proof. Since the proof closely parallels that of Theorem 5.4.1, it is only sketched here. There exists a C such that (P_i, C) is stable for $i = 0, \cdots, l$ if and only if there exist matrices $R_i \in \mathbf{M}(\mathbf{S})$ for $i = 1, \cdots, l$ and $U_i \in \mathbf{U}(\mathbf{S})$ for $i = 0, \cdots, l$ such that

$$[I \ R_0] \begin{bmatrix} Y_0 & X_0 \\ -\tilde{N}_0 & \tilde{D}_0 \end{bmatrix} = U_i [I \ R_i] \begin{bmatrix} Y_i & X_i \\ -\tilde{N}_i & \tilde{D}_i \end{bmatrix}, \quad i = 1, \cdots, l, \tag{5.4.19}$$

(see (5.4.7)), or equivalently,

$$[I \ R_0] \begin{bmatrix} Y_0 & X_0 \\ -\tilde{N}_0 & \tilde{D}_0 \end{bmatrix} \begin{bmatrix} D_i & -\tilde{X}_i \\ N_i & \tilde{Y}_i \end{bmatrix} = U_i [I \ R_i], \quad i = 1, \cdots, l. \tag{5.4.20}$$

The first of these equations gives

$$A_i + R_0 B_i = U_i, \quad i = 1, \cdots, l, \tag{5.4.21}$$

while the second can be written as

$$R_i = U_i^{-1} [I \ R_0] \begin{bmatrix} Y_0 & X_0 \\ -\tilde{N}_0 & \tilde{D}_0 \end{bmatrix} \begin{bmatrix} -\tilde{X}_i \\ \tilde{Y}_i \end{bmatrix}, \quad i = 1, \cdots, l. \tag{5.4.22}$$

It is left to the reader to complete the proof.   □

See Section 7.6 for some results concerning the genericity of simultaneous stabilizability.

Finally, the problem of simultaneously stabilizing l plants $P_1, \cdots, P_l$ using a stable compensator is equivalent to the problem of simultaneously stabilizing $l + 1$ plants $P_0 = 0, P_1, \cdots, P_l$.

PROBLEMS

5.4.1. Show that A and B in (5.4.2) are right-coprime.

5.4.2. Prove Corollary 5.4.3 from Theorem 5.4.1. (Hint: If P_0 is strictly proper and stable, then one may take

$$\begin{bmatrix} Y_0 & X_0 \\ -\tilde{N}_0 & \tilde{D}_0 \end{bmatrix} = \begin{bmatrix} I & 0 \\ -P_0 & I \end{bmatrix}.$$

Now apply the theorem.)

5.5   MULTI-COMPENSATOR CONFIGURATION

The problem studied in this section is in a sense the dual of the simultaneous stabilization problem studied in the previous section. In the latter, the emphasis is on determining whether there exists a single compensator that stabilizes each of two (or more) plants. In contrast, the objective of this section is to determine conditions under which a single plant is stabilized by two compensators, each acting alone or both acting together.

To state the problem precisely, consider the system shown in Figure 5.3, where $P, C_1, C_2 \in$ $\mathbf{M}(\mathbb{R}(s))$ and have compatible dimensions. The relationship between $e = (e_1, e_2, e_3)$ and $u = (u_1, u_2, u_3)$ is given by

$$\begin{bmatrix} e_1 \\ e_2 \\ e_3 \end{bmatrix} = H(P, C_1, C_2) \begin{bmatrix} u_1 \\ u_2 \\ u_3 \end{bmatrix} , \tag{5.5.1}$$

where, letting C denote $C_1 + C_2$,

$$H(P, C_1, C_2) = \begin{bmatrix} (I + PC)^{-1} & -PC_2(I + PC)^{-1} & -P(I + CP)^{-1} \\ -PC_1(I + PC)^{-1} & (I + PC)^{-1} & -P(I + CP)^{-1} \\ C_1(I + PC)^{-1} & C_2(I + PC)^{-1} & (I + CP)^{-1} \end{bmatrix} . \tag{5.5.2}$$

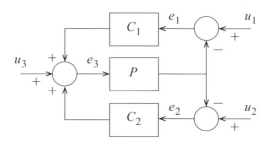

Figure 5.3: Multi-Compensator Configuration.

The *Reliable Stabilization Problem* (RSP) investigated in this section can now be stated precisely: Given P, find C_1 and C_2 such that

(i) $H(P, C_1, C_2) \in \mathbf{M}(\mathbf{S})$,

(ii) $H(P, C_1, 0) \in \mathbf{M}(\mathbf{S})$,

(iii) $H(P, 0, C_2) \in \mathbf{M}(\mathbf{S})$.

If such C_1, C_2 can be found, then the system of Figure 5.3 is externally and internally stable whenever both controllers C_1 and C_2 are in operation, as well as when only one (either one) of the controllers is operational and the other is set to zero. Such a system becomes unstable only when *both* controllers fail simultaneously, and is therefore reliable against a single controller failure.

Comparing the present problem with that in Section 5.4, we see that in the latter case the objective is to maintain stability in the face of possible structural changes in the plant, whereas here the objective is to guard against possible compensator failure. In a practical problem, the nature and location of the expected failures would dictate which mathematical formalism is more appropriate.

It is easy to see from (5.5.2) that $H(P, C_1, 0) \in \mathbf{M}(\mathbf{S})$ if and only if $C_1 \in S(P)$; similarly, $H(P, 0, C_2) \in \mathbf{M}(\mathbf{S})$ if and only if $C_2 \in S(P)$. However, $H(P, C_1, C_2) \in \mathbf{M}(\mathbf{S})$ implies, but is not necessarily implied by $C_1 + C_2 \in S(P)$. The reason is that if $H(P, C_1, C_2) \in \mathbf{M(S)}$, then

$$
\begin{aligned}
(I + PC)^{-1} &= H_{11} \in \mathbf{M(S)} , \\
(I + CP)^{-1} &= H_{33} \in \mathbf{M(S)} , \\
P(I + CP)^{-1} &= -H_{13} \in \mathbf{M(S)} , \\
C(I + PC)^{-1} &= H_{31} + H_{32} \in \mathbf{M(S)} .
\end{aligned}
\tag{5.5.3}
$$

As a result, $C \in S(P)$. However, $H_{31} + H_{32} \in \mathbf{M(S)}$ does not necessarily imply that H_{31} and H_{32} individually belong to $\mathbf{M(S)}$. Therefore $C_1 + C_2 \in S(P)$ does not necessarily imply that $H(P, C_1, C_2) \in \mathbf{M(S)}$. Hence, the RSP is not quite the same thing as the modified RSP, which can be stated as follows: Given a plant P, find $C_1, C_2 \in S(P)$ such that $C_1 + C_2 \in S(P)$. Nevertheless, we shall find it profitable to approach the RSP via the modified RSP.

We present at once the first main result of this section.

Theorem 5.5.1   Suppose $P \in \mathbf{M}(\mathbb{R}(s))$ and $C_1 \in S(P)$. Then there exists a $C_2 \in S(P)$ such that C_1 and C_2 together solve the Reliable Stabilization Problem.

In other words, not only does the RSP have a solution for *every* plant P, but *one* of the two stabilizing controllers can be arbitrarily specified.

Theorem 5.5.2   Suppose $P \in \mathbf{M}(\mathbb{R}(s))$. Then there exists a $C \in S(P)$ such that $C_1 = C_2 = C$ solve the RSP.

This theorem shows that it is always possible to find a solution to the RSP in the form of a "duplicate" controller, in the sense that $C_1 = C_2$. It should be noted however that the multi-controller configuration of Figure 5.3 is *not* the same as having a back-up controller; rather, in the present set-up, *both* controllers are connected at all times.

The proof of Theorem 5.5.2 requires the following lemma.

Lemma 5.5.3   *Suppose $A, B \in \mathbf{M}(\mathbb{R}(s))$, and that the products AB, BA are both defined (and hence square). Then there exists a $Q \in \mathbf{M(S)}$ such that $I - AB + QBAB$ is unimodular.*

This lemma means that, whatever be the matrices A, B, the plant $BAB(I - AB)^{-1}$ is strongly stabilizable.

Proof.   Recall that the norm on $\mathbf{M(S)}$ is defined by

$$
\|F\| = \sup_{\omega} \bar{\sigma}(F(j\omega)), \ \forall F \in \mathbf{M(S)} ,
\tag{5.5.4}
$$

where $\bar{\sigma}(\cdot)$ denotes the largest singular value of a matrix. Then $I + F$ is unimodular whenever $\|F\| < 1$. In particular, $I - rAB$ is unimodular whenever $|r| < \|AB\|^{-1}$. Let k be an integer larger than $\|AB\|$. Then $(I - k^{-1}AB)$ is unimodular, and so is $(I - k^{-1}AB)^k$. By the binomial expansion,

$$(I - k^{-1}AB)^k = I - AB + \sum_{i=2}^{k} f_i(AB)^i , \tag{5.5.5}$$

where the f_i are appropriate real numbers. Now define

$$Q = \sum_{i=0}^{k-2} f_{i+2}(AB)^i A \in \mathbf{M}(\mathbf{S}) . \tag{5.5.6}$$

Then clearly

$$I - AB + QBAB = (I - k^{-1}AB)^k , \tag{5.5.7}$$

is unimodular. $\square$

Proof of Theorem 5.5.1. Suppose $P \in \mathbf{M}(\mathbb{R}(s))$ and $C_1 \in S(P)$ are specified. Let (N, D), $(\tilde{D}, \tilde{N})$ be any right-coprime factorization and left-coprime factorization, respectively, of P over $\mathbf{M}(\mathbf{S})$. The fact that $C_1 \in S(P)$ implies, by Theorem 5.2.1, that $C_1 = Y^{-1}X = \tilde{X}\tilde{Y}^{-1}$, where $X, Y, \tilde{X}, \tilde{Y} \in \mathbf{M}(\mathbf{S})$ satisfy $XN + YD = I$, $\tilde{N}\tilde{X} + \tilde{D}\tilde{Y} = I$. Moreover, $Y^{-1}X = \tilde{X}\tilde{Y}^{-1}$ implies that $Y\tilde{X} = X\tilde{Y}$.

Using Lemma 5.5.3, select a matrix $Q \in \mathbf{M}(\mathbf{S})$ such that $I - XN + QNXN$ is unimodular. From Theorem 5.2.1, the controller C defined by

$$C = (Y - Q\tilde{Y}\tilde{N})^{-1}(X + Q\tilde{Y}\tilde{D}) , \tag{5.5.8}$$

is in $S(P)$. Let $C_2 = C - C_1$. We now show that C_2 is also in $S(P)$, which shows that C_1 and C_2 together solve the modified RSP. Now

$$\begin{aligned}
C_2 = C - C_1 &= (Y - Q\tilde{Y}\tilde{N})^{-1}(X + Q\tilde{Y}\tilde{D}) - \tilde{X}\tilde{Y}^{-1} \\
&= (Y - Q\tilde{Y}\tilde{N})^{-1}[(X + Q\tilde{Y}\tilde{D})\tilde{Y} - (Y - Q\tilde{Y}\tilde{N})\tilde{X}]\tilde{Y}^{-1} \\
&= (Y - Q\tilde{Y}\tilde{N})^{-1}[X\tilde{Y} - Y\tilde{X} + Q\tilde{Y}(\tilde{D}\tilde{Y} + \tilde{N}\tilde{X})]\tilde{Y}^{-1} \\
&= (Y - Q\tilde{Y}\tilde{N})^{-1}Q\tilde{Y}\tilde{Y}^{-1}, \ \text{by (5.5.4) and (5.5.5)} \\
&= (Y - Q\tilde{Y}\tilde{N})^{-1}Q \\
&= \tilde{D}_{c_2}^{-1}\tilde{N}_{c_2}, \tag{5.5.9}
\end{aligned}$$

where $\tilde{D}_{c_2} = Y - Q\tilde{Y}\tilde{N}$, $\tilde{N}_{c_2} = Q$. At this stage, it has not been shown that $\tilde{D}_{c_2}$, $\tilde{N}_{c_2}$ are left-coprime. But let us anyway compute the "return difference" matrix $\tilde{D}_{c_2}D + \tilde{N}_{c_2}N$. This gives

$$\begin{aligned}
\tilde{D}_{c_2}D + \tilde{N}_{c_2}N &= (Y - Q\tilde{Y}\tilde{N})D + QN \\
&= YD - Q\tilde{Y}\tilde{N}D + QN \\
&= YD - Q\tilde{Y}\tilde{D}N + QN, \ \text{since } \tilde{N}D = \tilde{D}N . \tag{5.5.10}
\end{aligned}$$

Next, note that

$$\begin{bmatrix} Y & X \\ -\tilde{N} & \tilde{D} \end{bmatrix} \begin{bmatrix} D & -\tilde{X} \\ N & \tilde{Y} \end{bmatrix} = I . \tag{5.5.11}$$

Thus, the two matrices in (5.5.11) are the inverses of each other. Hence, their product in the opposite order is also equal to the identity matrix; that is,

$$\begin{bmatrix} D & -\tilde{X} \\ N & \tilde{Y} \end{bmatrix} \begin{bmatrix} Y & X \\ -\tilde{N} & \tilde{D} \end{bmatrix} = I . \tag{5.5.12}$$

In particular, $NX + \tilde{Y}\tilde{D} = I$, so that $\tilde{Y}\tilde{D} = I - NX$; similarly, $YD = I - XN$. Substituting these identities in (5.5.10) gives

$$\tilde{D}_{c_2}D + \tilde{N}_{c_2}N = I - XN - QN - QNXN + QN$$
$$= I - XN + QNXN , \tag{5.5.13}$$

which is unimodular by construction. Hence, $C_2 \in S(P)$. This also shows, *a fortiori*, the left-coprimeness of $(\tilde{D}_{c_2}, \tilde{N}_{c_2}) = (Y - Q\tilde{Y}\tilde{N}, Q)$.

To complete the proof, it is necessary to show that the transfer matrix $H(P, C_1, C_2) \in \mathbf{M(S)}$. For this purpose, it is more convenient to work with the quantity $K(P, C_1, C_2)$, which is the transfer matrix between $(u_1, u_3, u_1 - u_2)$ and $(e_1, e_3, e_1 - e_2)$. Routine calculations show that

$$\begin{bmatrix} e_1 \\ e_3 \\ e_1 - e_2 \end{bmatrix} = \begin{bmatrix} u_1 \\ u_3 \\ u_1 - u_3 \end{bmatrix} - \begin{bmatrix} 0 & P & 0 \\ -(C_1 + C_2) & 0 & C_2 \\ 0 & 0 & 0 \end{bmatrix} \begin{bmatrix} e_1 \\ e_3 \\ e_1 - e_2 \end{bmatrix} . \tag{5.5.14}$$

Since $K(P, C_1, C_2)$ is obtained by pre- and post-multiplying $H(P, C_1, C_2)$ by unimodular (and in fact nonsingular constant) matrices, it follows that $H(P, C_1, C_2) \in \mathbf{M(S)}$ if and only if $K(P, C_1, C_2) \in \mathbf{M(S)}$. Now from (5.5.14),

$$K(P, C_1, C_2) = \begin{bmatrix} I & P & 0 \\ -(C_1 + C_2) & I & C_2 \\ 0 & 0 & I \end{bmatrix}^{-1} . \tag{5.5.15}$$

The matrix in (5.5.15) is block-lower triangular, which makes it easy to invert. Also, by construction,

$$H(P, C_1 + C_2) = H(P, C)$$
$$= \begin{bmatrix} I & -P \\ C_1 + C_2 & I \end{bmatrix}^{-1}$$
$$= \begin{bmatrix} I - N(X + Q\tilde{Y}\tilde{D}) & -N(Y - Q\tilde{Y}\tilde{N}) \\ D(X + Q\tilde{Y}\tilde{D}) & D(Y - Q\tilde{Y}\tilde{N}) \end{bmatrix} . \tag{5.5.16}$$

It now follows from (5.5.15) and (5.5.16) that

$$K(P, C_1, C_2) = \begin{bmatrix} I - N(X + Q\tilde{Y}\tilde{D}) & -N(Y - Q\tilde{Y}\tilde{N}) & -NQ \\ D(X + Q\tilde{Y}\tilde{D}) & D(Y - Q\tilde{Y}\tilde{N}) & DQ \\ 0 & 0 & I \end{bmatrix}, \tag{5.5.17}$$

which belongs to $\mathbf{M(S)}$. Therefore C_1 and C_2 solve the RSP. $\qquad\square$

The above method of proof makes it clear that, whenever $\tilde{D}_{c_2}$ divides $\tilde{D}_c$, we have that $H(P, C) \in \mathbf{M(S)}$ implies that $H(P, C_1, C_2) \in \mathbf{M(S)}$. Thus, in order to prove Theorem 5.5.2, it is enough to prove the existence of a controller C such that $C, 2C \in S(P)$. The proof of Theorem 5.5.2 depends on the following lemma.

Lemma 5.5.4   *Given a plant $P \in \mathbf{M}(\mathbb{R}(s))$, let (N, D), $(\tilde{D}, \tilde{N})$ be any r.c.f. and l.c.f. of P, and let $X, Y \in \mathbf{M(S)}$ be solutions of $XN + YD = I$. Then there exists an $R \in \mathbf{M(S)}$ such that $I + XN + R\tilde{D}N$ is unimodular.*

Proof.   It is first shown that the matrices $I + XN$, $\tilde{D}N$ are right-coprime. From Theorem 4.1.16, one can select matrices $\tilde{X}, \tilde{Y} \in \mathbf{M(S)}$ such that (5.5.11) and (5.5.12) hold. Suppose now that M is a right divisor of both $I + XN$ and $\tilde{D}N$, denoted by $M|(I + XN)$, $M|\tilde{D}N$. This implies, successively, that

$$\begin{aligned} &M|\tilde{Y}\tilde{D}N, \; M|(I - NX)N, \text{ since } NX + \tilde{Y}\tilde{D} = I \;, \\ &M|N(I - XN), \; M|(I + NX)N \;, \\ &M|N \text{ since } N = [(I + NX)N + (I - NX)N]/2 \;, \\ &M|XN \;, \\ &M|I \text{ since } M|(I + XN), \; M|XN \;. \end{aligned} \tag{5.5.18}$$

This last step shows that M is unimodular.

Now let C_{+e} denote the extended closed right half-plane, i.e., $C_{+e} = \{s : \mathrm{Re}\; s \geq 0\} \bigcup \{\infty\}$. The next step is to show that $|I + X(s)N(s)| > 0$ whenever $s \in C_{+e}$ is real and $\tilde{D}(s)N(s) = 0$. It would then follow from Theorem 5.3.1 that $I + XN + R\tilde{D}N$ is unimodular for some $R \in \mathbf{M(S)}$. Suppose $(\tilde{D}N)(s) = 0$. Then

$$\begin{aligned} (\tilde{Y}\tilde{D}N)(s) = 0 &\Longrightarrow [(I - NX)N](s) = 0 \\ &\Longrightarrow N(s) = NXN(s) \\ &\Longrightarrow XN(s) = XNXN(s) = [XN(s)]^2 \;. \end{aligned} \tag{5.5.19}$$

Let $\alpha = \sqrt{2} - 1 \approx .414$. Then it is easy to verify that $1 - 2\alpha = \alpha^2$. Thus,

$$\begin{aligned} I + XN(s) &= I + 2\alpha XN(s) + \alpha^2(XN)(s) \\ &= I + 2\alpha(XN)(s) + \alpha^2[(XN)(s)]^2, \text{ by } (5.5.18) \\ &= [I + \alpha(XN)(s)]^2 \;. \end{aligned} \tag{5.5.20}$$

$$|I + (XN)(s)| = |I + \alpha(XN)(s)|^2 \geq 0 \;. \tag{5.5.21}$$

However, since $I + XN$ and $\tilde{D}N$ are right-coprime, the smallest invariant factor of $\tilde{D}N$ and $|I + XN|$ are coprime. Hence, $|I + (XN)(s)| \neq 0$, which implies in conjunction with (5.5.21) that $|I + (XN)(s)| > 0$.   □

Proof of Theorem 5.5.2. Let $C = (Y - R\tilde{N})^{-1}(X + R\tilde{D})$. Then $2C = (Y - R\tilde{N})^{-1} \cdot 2(X + R\tilde{D})$. Clearly $C \in S(P)$. The return difference matrix corresponding to P and $2C$ is

$$(Y - R\tilde{N})D + 2 \cdot (X + R\tilde{D})N = I + XN + R\tilde{D}N , \qquad (5.5.22)$$

which is unimodular by construction. Thus, $2C \in S(P)$. By earlier remarks, this is enough to show that $C_1 = C_2 = C$ solve the RSP.   □

We summarize the results presented up to now by providing two algorithms for reliable stabilization.

Given a plant P and a controller $C_1 \in S(P)$, to find another controller C_2 which, together with C_1, reliably stabilizes P.

Step 1. Find r.c.f.'s (N, D) of P and $(\tilde{X}, \tilde{Y})$ of C_1 and l.c.f.'s $(\tilde{D}, \tilde{N})$ of P and (Y, X) of C_1 such that $XN + YD = I$ and $\tilde{N}\tilde{X} + \tilde{D}\tilde{Y} = I$ are satisfied.

Step 2. Find a Q such that $I - XN + QNXN$ is unimodular, either by using the procedure described in the proof of Lemma 5.5.3, or by any other means.

Step 3. Let $C_2 = (Y - Q\tilde{Y}\tilde{N})^{-1}Q$.

Given a plant P, to find a "duplicate" controller C such that $C_1 = C_2 = C$ solves the reliable stabilization problem.

Step 1. Find any r.c.f. (N, D) and any l.c.f. $(\tilde{D}, \tilde{N})$ of P, together with a particular solution (X, Y) of $XN + YD = I$.

Step 2. Find an $R \in \mathbf{M(S)}$ such that $I + XN + R\tilde{D}N$ is unimodular.

Step 3. Let $C = (Y - R\tilde{N})^{-1}(X + R\tilde{D})$.

Example 5.5.5 Suppose it is desired to reliably stabilize the plant

$$p(s) = \frac{s^2 - 1}{s^2 - 4} .$$

Since this plant is nonminimum phase, robustness recovery techniques cannot be used to solve the problem. To use Algorithm 5.5 express p in the form $p = n/d$, where

$$n(s) = \frac{s - 1}{s + 2}, \quad d(s) = \frac{s - 2}{s + 1} .$$

A particular solution to the Bezout identity $xn + yd = 1$ is given by

$$x(s) = \frac{3(s+2)}{s+1}, \quad y(s) = -2 .$$

Hence, the controller

$$c_1 = \frac{x}{y} = -1.5\frac{s+2}{s+1} ,$$

stabilizes p. We shall now find another controller c_2 which, together with c_1, reliably stabilizes p. The first step is to find a q such that $1 - xn + qnxn$ is a unit. This can be done using the method of Lemma 5.5.3. Note that

$$(xn)(s) = 3\frac{s-1}{s+1} ,$$

so that $\|xn\| = 3$. Hence, $(1 - 0.25xn)^4$ is a unit, and is of the form $1 + xn + qnxn$. In fact a routine computation yields that

$$q = x \left\{ \frac{3}{8} - \frac{xn}{16} + \frac{xn^2}{256} \right\} .$$

The expression for $c_2 = q/(y - yqn) = q/y(1 - qn)$ is easily obtained by using the variable $z = (s - 1)/(s + 1)$, since in this case xn is just $3z$. This leads to

$$c_2 = \frac{3(z-2)(9z^2 - 48z + 96)}{4(-27z^3 + 144z^2 - 288z + 256)} ,$$

which is a third-order controller.

5.6   TWO-PARAMETER COMPENSATORS

In this section, the feedback system shown in Figure 5.1 is replaced by a more general type of system, which employs what is generally referred to as a two-degrees-of-freedom compensator. For reasons that will become clear shortly, it is referred to here as a *two-parameter compensator*.

Suppose a plant P is specified, and it is desired to stabilize P using a feedback compensator. If e and y denote the plant input and output, respectively, and u denotes the external input, then the *most general* feedback compensation scheme is given by

$$e = C(u, y) , \tag{5.6.1}$$

where C is a (possibly nonlinear) operator. The most general linear time-invariant dynamical scheme is

$$e = C_1u - C_2y , \tag{5.6.2}$$

where $C_1, C_2 \in \mathbf{M}(\mathbb{R}(s))$ and have the same number of rows.[2] If $C_1 = C_2 = C$, then (5.6.2) represents the feedback configuration of Figure 5.1, but otherwise the compensation law (5.6.2) is more general. For this reason, the compensator (5.6.2), which can also be written as

$$e = [C_1 \quad -C_2] \begin{bmatrix} u \\ y \end{bmatrix}, \tag{5.6.3}$$

is referred to as a two-parameter compensator. It is of course understood that the two inputs u and y can themselves be vectors. It is worth emphasizing again that (5.6.3) represents the most general feedback compensation scheme involving the external input and the plant output. Thus, there is no "three-parameter" compensator, for example, since *all* external inputs are included in u.

One way in which the general feedback law (5.6.2) can be implemented is shown in Figure 5.4. As shown, this implementation makes no sense unless C_1 is stable: If C_1 is unstable, then certain bounded inputs u will produce unbounded outputs $C_1 u$, and as a result the system can never be internally stable. Thus, we search for an alternate implementation of (5.6.2) that avoids this problem.

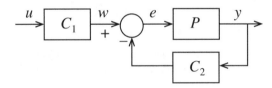

Figure 5.4: Infeasible Implementation of a Two-Parameter Compensator.

To arrive at a feasible implementation of the two-parameter compensator, define $C = [C_1 \ C_2]$, and let $(\tilde{D}_c, [\tilde{N}_{c_1} \ \tilde{N}_{c_2}])$ be an l.c.f. of C. Thus, $C_1 = \tilde{D}_c^{-1}\tilde{N}_{c_1}$ and $C_2 = \tilde{D}_c^{-1}\tilde{N}_{c_2}$. However, the left-coprimeness of $\tilde{D}_c$ and $[\tilde{N}_{c_1} \ \tilde{N}_{c_2}]$ does *not* imply that $\tilde{D}_c$ and $\tilde{N}_{c_1}$ are left-coprime; i.e., $(\tilde{D}_c, \tilde{N}_{c_1})$ need not be an l.c.f. of C_1. Similar remarks apply to $\tilde{D}_c, \tilde{N}_{c_2}$. In fact, suppose $(\tilde{D}_1, \tilde{N}_1)$ (resp. $(\tilde{D}_2, \tilde{N}_2)$) is an l.c.f. of C_1 (resp. C_2). Then $\tilde{D}_c$ is a least common left multiple of $\tilde{D}_1$ and $\tilde{D}_2$; moreover, if $\tilde{D}_c = \tilde{A}\tilde{D}_1 = \tilde{B}\tilde{D}_2$, then $[\tilde{N}_{c_1} \ \tilde{N}_{c_2}] = [\tilde{A}\tilde{N}_1 \ \tilde{B}\tilde{N}_2]$ (see Problem 5.6.1).

Once $C = [C_1 \ C_2]$ has been factorized as $\tilde{D}_c^{-1}[\tilde{N}_{c_1} \ \tilde{N}_{c_2}]$, the feedback law (5.6.2) can be implemented as shown in Figure 5.5. In this figure, it is assumed that the plant has been factored as $N_p D_p^{-1}$ where $N_p, D_p \in \mathbf{M}(\mathbf{S})$ are right-coprime. Further, exogenous signals u_2 and u_3 are included to model the presence of errors in measuring the outputs of C and P, respectively.

The system of Figure 5.5 is *stable* if the matrix W mapping (u_1, u_2, u_3) into (y_1, y_2) belongs to $\mathbf{M}(\mathbf{S})$. It can be shown that this is equivalent to requiring the matrix mapping (u_1, u_2, u_3) into (z_1, z_2, y_1, y_2) to belong to $\mathbf{M}(\mathbf{S})$ (see Problem 5.6.2). Thus, the present notion of stability corresponds to requiring all "internal" quantities z_1, z_2, y_1, y_2 to belong to $\mathbf{M}(\mathbf{S})$ whenever

---

[2]The minus sign before C_2 is to represent negative feedback from y to e.

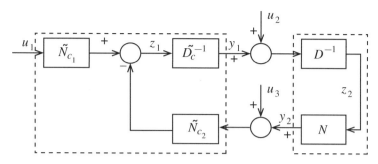

Figure 5.5: Feasible Implementation of a Two-Parameter Compensator.

$u_1, u_2, u_3 \in \mathbf{M(S)}$. Lemma 5.6.1 below gives necessary and sufficient conditions for the system of Figure 5.5 to be stable, in terms of the factorizations of C and P.

Lemma 5.6.1 *Let (N_p, D_p) be a r.c.f. of P and let $(\tilde{D}_c, [\tilde{N}_{c_1} \ \ \tilde{N}_{c_2}])$ be an l.c.f. of $C = [C_1 \ \ C_2]$. Then the system of Figure 5.5 is stable if and only if*

$$\Delta := \tilde{D}_c D_p + \tilde{N}_{c_2} N_p \in \mathbf{U(S)} . \tag{5.6.4}$$

Remarks 5.6.2   1) Clearly a *necessary* condition for Δ to be unimodular is that $\tilde{D}_c$ and $\tilde{N}_{c_2}$ be left-coprime. In view of earlier discussion, this can be interpreted as follows: As before, let $(\tilde{D}_1, \tilde{N}_1), (\tilde{D}_2, \tilde{N}_2)$ be l.c.f.'s of C_1 and C_2 respectively, and suppose $\tilde{D}_c = \tilde{A}\tilde{D}_1 = \tilde{B}\tilde{D}_2$ is a least common left multiple of $\tilde{D}_1$ and $\tilde{D}_2$. Then, since $\tilde{N}_{c_2} = \tilde{B}\tilde{N}_2$, we see that $\tilde{D}_c = \tilde{B}\tilde{D}_2, \tilde{N}_{c_2} = \tilde{B}\tilde{N}_2$ are left-coprime (if and) only if $\tilde{B}$ is unimodular. Therefore, $\tilde{D}_c = \tilde{D}_2$ must be a least common left multiple of $\tilde{D}_1, \tilde{D}_2$; in other words, $\tilde{D}_2$ is a left multiple of $\tilde{D}_1$. To summarize, the two-parameter compensator (5.6.3) cannot stabilize P unless the left denominator of C_1 divides the left denominator of C_2. This means that every C_{+e}-pole of C_1 must also be a pole of C_2 of at least the same McMillan degree.

Proof. The equations describing the system of Figure 5.5 can be written in terms of the quantities y_1 and z_2, which are the outputs of the "inversion" elements $\tilde{D}_c^{-1}$ and D_p^{-1}, respectively. Thus,

$$e_2 = D_p z_2 = y_1 + u_2 \implies y_1 - D_p z_2 = -u_2 , \tag{5.6.5}$$

$$z_1 = \tilde{D}_c y_1 = \tilde{N}_{c_1} u_1 - \tilde{N}_{c_2}(u_3 + y_2) = \tilde{N}_{c_1} u_1 - \tilde{N}_{c_2} u_3 - \tilde{N}_{c_2} N_p z_2$$

$$\implies \tilde{D}_c y_1 + \tilde{N}_{c_2} N_p z_2 = \tilde{N}_{c_1} u_1 - \tilde{N}_{c_2} u_3 . \tag{5.6.6}$$

Equations (5.6.5) and (5.6.6) can be combined in the form

$$
\begin{bmatrix} I & -D_p \\ \tilde{D}_c & \tilde{N}_{c_2} N_p \end{bmatrix} \begin{bmatrix} y_1 \\ z_2 \end{bmatrix} = \begin{bmatrix} 0 & -I & 0 \\ \tilde{N}_{c_1} & 0 & -\tilde{N}_{c_2} \end{bmatrix} \begin{bmatrix} u_1 \\ u_2 \\ u_3 \end{bmatrix}. \tag{5.6.7}
$$

Now the outputs y_1 and y_2 are given by

$$
\begin{bmatrix} y_1 \\ y_2 \end{bmatrix} = \begin{bmatrix} I & 0 \\ 0 & N_p \end{bmatrix} \begin{bmatrix} y_1 \\ z_2 \end{bmatrix}. \tag{5.6.8}
$$

It is immediate from (5.6.7) and (5.6.8) that

$$
\begin{bmatrix} y_1 \\ y_2 \end{bmatrix} = N_r D^{-1} \tilde{N}_l \begin{bmatrix} u_1 \\ u_2 \\ u_3 \end{bmatrix}, \tag{5.6.9}
$$

where

$$
N_r = \begin{bmatrix} I & 0 \\ 0 & N_p \end{bmatrix}, \quad D = \begin{bmatrix} I & -D_p \\ \tilde{D}_c & \tilde{N}_{c_2} N_p \end{bmatrix}, \quad \tilde{N}_l = \begin{bmatrix} 0 & -I & 0 \\ \tilde{N}_{c_1} & 0 & \tilde{N}_{c_2} \end{bmatrix}. \tag{5.6.10}
$$

The next step is to show that N_r, D are right-coprime, and that $\tilde{N}_l$, D are left-coprime. For this purpose, select $X_p, Y_p, \tilde{X}_{c_1}, \tilde{X}_{c_2}, \tilde{Y}_c \in \mathbf{M(S)}$ such that

$$
X_p N_p + Y_p D_p = I, \quad [\tilde{N}_{c_1} \ \ \tilde{N}_{c_2}] \begin{bmatrix} \tilde{X}_{c_1} \\ \tilde{X}_{c_2} \end{bmatrix} + \tilde{D}_c \tilde{Y}_c = I. \tag{5.6.11}
$$

Then, recalling that $\Delta = \tilde{D}_c D_p + \tilde{N}_{c_2} N_p$, it is routine to verify that

$$
\begin{bmatrix} \Delta Y_p + I & \Delta X_p \\ Y_p & X_p \end{bmatrix} \begin{bmatrix} I & 0 \\ 0 & N_p \end{bmatrix} + \begin{bmatrix} \tilde{D}_c - \Delta Y_p & -I \\ -Y_p & 0 \end{bmatrix} \begin{bmatrix} I & -D_p \\ \tilde{D}_c & \tilde{N}_{c_2} N_p \end{bmatrix} = I, \tag{5.6.12}
$$

$$
\begin{bmatrix} 0 & -I & 0 \\ \tilde{N}_{c_1} & 0 & -\tilde{N}_{c_2} \end{bmatrix} \begin{bmatrix} \tilde{X}_{c_1} \Delta & \tilde{X}_{c_1} \\ \tilde{Y}_c \Delta - I & \tilde{Y}_c \\ -\tilde{X}_{c_2} \Delta & -\tilde{X}_{c_2} \end{bmatrix} + \begin{bmatrix} I & -D_p \\ \tilde{D}_c & \tilde{N}_c N_p \end{bmatrix} \begin{bmatrix} \tilde{Y}_c \Delta - D_p & \tilde{Y}_c \\ -I & 0 \end{bmatrix} = I. \tag{5.6.13}
$$

By definition, the system under study is stable if and only if $N_r D^{-1} \tilde{N}_l \in \mathbf{M(S)}$. By Corollary 4.3.9, this is so if and only if $D \in \mathbf{U(S)}$. The proof is completed by showing that $D \in \mathbf{U(S)}$ if and only if $\Delta \in \mathbf{U(S)}$. From (5.6.10), we see that Δ is just the Schur complement of D with respect to the identity matrix in the upper left-hand corner. Hence, from Fact B.1.9, we get $|D| = |I| \cdot |\Delta| = |\Delta|$. Therefore $D \in \mathbf{U(S)}$ if and only if $\Delta \in \mathbf{U(S)}$ (if and only if $|D| = |\Delta|$ is a unit of $\mathbf{S}$). $\square$

Now we come to the main result of this section.

Theorem 5.6.3   Let P be a given plant. Suppose (N_p, D_p), $(\tilde{D}_p, \tilde{N}_p)$ are any r.c.f. and any l.c.f. of P, and that $X, Y \in \mathbf{M}(\mathbf{S})$ satisfy $XN_p + YD_p = I$. Then the set of all two-parameter compensators that stabilize P is given by

$$S_2(P) = \{(Y - R\tilde{N}_p)^{-1}[Q \ \ X + R\tilde{D}_p]\}, \tag{5.6.14}$$

where $Q \in \mathbf{M}(\mathbf{S})$ is arbitrary, and $R \in \mathbf{M}(\mathbf{S})$ is arbitrary but for the constraint that $|Y - R\tilde{N}_p| \neq 0$. The set of all possible stable transfer matrices from (u_1, u_2, u_3) to (y_1, y_2) is the set of all matrices of the form

$$\begin{bmatrix} D_p Q & D_p(Y - R\tilde{N}_p) - I & -D_p(X + R\tilde{D}_p) \\ N_p Q & N_p(Y - R\tilde{N}_p) & -N_p(X + R\tilde{D}_p) \end{bmatrix}$$

$$= \begin{bmatrix} D_p \\ N_p \end{bmatrix} [Q \ R] \begin{bmatrix} I & 0 & 0 \\ 0 & -\tilde{N}_p & -\tilde{D}_p \end{bmatrix} + \begin{bmatrix} 0 & D_p Y - I & -D_p X \\ 0 & N_p Y & -N_p X \end{bmatrix}, \tag{5.6.15}$$

where $Q, R \in \mathbf{M}(\mathbf{S})$ are arbitrary but for the constraint that $|Y - R\tilde{N}_p| \neq 0$.

Proof. Based on the results of Section 5.2, specifically Theorem 5.2.1, it follows that the return difference matrix Δ of (5.6.4) is unimodular if and only if $\tilde{D}_c^{-1} \tilde{N}_{c_2}$ equals $(Y - R\tilde{N}_p)^{-1}(X + R\tilde{D}_p)$ for some $R \in \mathbf{M}(\mathbf{S})$ such that $|Y - R\tilde{N}_p| \neq 0$. Moreover, Δ is independent of $\tilde{N}_{c_1}$. This leads at once to (5.6.14). Now (5.6.15) follows by routine computation.   □

Several remarks are in order concerning this theorem.

1) Equation (5.6.15) shows that the transfer matrix from (u_1, u_2, u_3) to (y_1, y_2) involves *two independent* parameters Q and R. This is the most important feature of the two-parameter compensator (and it incidentally explains the name of the scheme). For comparison, recall that from (5.2.8) that the set of all stable transfer matrices from (u_1, u_2) to (y_1, y_2) in a one-parameter scheme, such as in Figure 5.1, is

$$\begin{bmatrix} D_p(X + R\tilde{D}_p) & D_p(Y - R\tilde{N}_p) - I \\ N_p(X + R\tilde{D}_p) & N_p(Y - R\tilde{N}_p) \end{bmatrix}. \tag{5.6.16}$$

Thus, the two-parameter scheme offers greater flexibility in that the transfer matrix from u_1 to the outputs can be adjusted "independently" of that between u_2 and the outputs. This is not so in the case of the one-parameter scheme. This flexibility is further illustrated in subsequent remarks.

2) Suppose P is a *stable* plant; then (P, I), (I, P) are respectively an r.c.f. and an l.c.f. of P, and one can take $X = 0$, $Y = I$. Then the set of all possible stable transfer matrices between u_1 and (y_1, y_2), subject to internal stability, is

$$\begin{bmatrix} Q \\ PQ \end{bmatrix}, \quad Q \in \mathbf{M}(\mathbf{S}), \tag{5.6.17}$$

in *both* schemes. Thus, in the case of stable plants, any stable transfer matrix between u_1 and the outputs that can be achieved using a two-parameter scheme can also be achieved using a one-parameter scheme. This is not true if P is unstable: If $\tilde{D}_p$ is not unimodular, then the set $\{D_p(X + R\tilde{D}_p) : R \in \mathbf{M}(\mathbf{S})\}$ is a *proper* subset of $\{D_pQ : Q \in \mathbf{M}(\mathbf{S})\}$. To see this, note that $D_pA = D_pB$ if and only if $A = B$, since $|D_p| \neq 0$. Thus,

$$\{D_p(X + R\tilde{D}_p) : R \in \mathbf{M}(\mathbf{S})\} = \{D_pQ : Q \in \mathbf{M}(\mathbf{S})\}$$
$$\Longleftrightarrow \{X + R\tilde{D}_p : R \in \mathbf{M}(\mathbf{S})\} = \mathbf{M}(\mathbf{S})$$
$$\Longleftrightarrow \{R\tilde{D}_p : R \in \mathbf{M}(\mathbf{S})\} = \mathbf{M}(\mathbf{S}) . \tag{5.6.18}$$

But this last equality holds if and only if $\tilde{D}_p$ is unimodular. Hence, if P is unstable, there exist stable transfer matrices between u_1 and the outputs that can be achieved with a two-parameter scheme but not a one-parameter scheme. In fact, (5.6.15) shows that by using a two-parameter scheme, the set of achievable transfer matrices between the external input u_1 and the plant output y_2 is just $\{N_pQ : Q \in \mathbf{M}(\mathbf{S})\}$, which is the same as if the plant were stable and equal to N_p. *Thus, in a two-parameter compensation scheme, the achievable performance between the external input and the plant output is not limited by the plant poles.* This point is discussed further in Section 6.7.

3) Let us think of u_2 and u_3 as sensor noises at the outputs of the compensator and plant, respectively. Then (5.6.15) and (5.6.16) show that, in terms of achievable performance between the sensor noises and the plant outputs, there is no difference between the two-parameter and the one-parameter schemes.

4) Suppose $M \in \mathbf{M}(\mathbf{S})$ is a right multiple of N_p. Then (5.6.15) shows that M can be realized as the transfer matrix between u_1 and y_2 by an appropriate two-parameter scheme. If N_p has full column rank, then there is a *unique* Q such that $M = N_pQ$. Nevertheless, there are *infinitely many* two-parameter schemes that realize the transfer matrix M, namely,

$$[C_1 \; C_2] = (Y - R\tilde{N}_p)^{-1}[Q \; X + R\tilde{D}_p] , \quad R \in \mathbf{M}(\mathbf{S}) . \tag{5.6.19}$$

In other words, specifying M fixes Q but not R. This "free" choice of R can be used to advantage in one of several ways: (i) It is possible *simultaneously* to achieve "good" transfer matrices $H_{y_2u_1}$ and $H_{y_2u_2}$, or $H_{y_2u_1}$ and $H_{y_2u_3}$. This is made quantitatively precise in Section 6.6. (ii) If P is strongly stabilizable, then *any* right multiple of N_p can be realized as $H_{y_2u_1}$ using only *stable* C_1 and C_2.

5) Problems such as strong stabilization, simultaneous stabilization, etc. are no different in the two-parameter case from the one-parameter case. This is because the "feedback" compensator C_2 has to stabilize P in the sense of Section 5.1 in order to achieve stability, irrespective of which scheme is used.

An important special case of the two-parameter compensator is the observer-controller configuration of Figure 5.6. In it, P is a given plant with an r.c.f. (N_p, D_p), and the matrices $X, Y \in \mathbf{M}(\mathbf{S})$ are chosen to satisfy $XN_p + YD_p = I$. The observer reconstructs the "internal state" z (sometimes referred to also as the "partial state"), and the controller feeds back z after multiplying it by M.

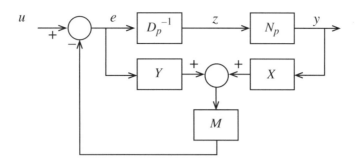

Figure 5.6: Observer-Controller Configuration.

To compare the observer-controller configuration of Figure 5.6 with the two-parameter compensator of Figure 5.5, let us redraw the former as shown in Figure 5.7. This results in a two-parameter compensator with

$$\tilde{N}_{C_1} = I \ , \quad \tilde{N}_{C_2} = MX \ , \quad \tilde{D}_c = I + MY \ . \tag{5.6.20}$$

Clearly $[\tilde{N}_{c_1} \ \tilde{N}_{c_2}]$ and $\tilde{D}_c$, are left-coprime, since

$$[I \ MX] \begin{bmatrix} I \\ 0 \end{bmatrix} + (I + MY) \cdot 0 = I \ . \tag{5.6.21}$$

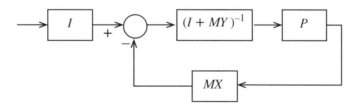

Figure 5.7: Rearrangement of the Observer-Controller Configuration.

Hence, from Lemma 5.6.1, this system is stable if and only if

$$\Delta = (I + MY)D_p + MXN_p = D_p + M \in \mathbf{U(S)} \ . \tag{5.6.22}$$

PROBLEMS

5.6.1. Suppose $(\tilde{D}_1, \tilde{N}_1)$, $(\tilde{D}_2, \tilde{N}_2)$ are l.c.f.'s of C_1 and C_2, respectively. Suppose $\tilde{D}_c = \tilde{A}\tilde{D}_1 = \tilde{B}\tilde{D}_2$ is a least common left multiple of $\tilde{D}_1$ and $\tilde{D}_2$. Show that $(\tilde{D}_c, [\tilde{A}\tilde{N}_1 \ \tilde{B}\tilde{N}_2])$ is an l.c.f. of $[C_1 \ C_2]$. (Hint: First show that $\tilde{A}, \tilde{B}$ are left-coprime.)

5.6.2. For the system of Figure 5.5, define the transfer matrices W and V by

$$\begin{bmatrix} z_1 \\ z_2 \\ y_1 \\ y_2 \end{bmatrix} = V \begin{bmatrix} u_1 \\ u_2 \\ u_3 \end{bmatrix}, \quad \begin{bmatrix} y_1 \\ y_2 \end{bmatrix} = W \begin{bmatrix} u_1 \\ u_2 \\ u_3 \end{bmatrix}.$$

Show that $W \in \mathbf{M(S)}$ if and only if $V \in \mathbf{M(S)}$.

5.7   REGULATION AND DECOUPLING

In this section, the focus is on control problems that can be reduced to the solvability of a set of linear equations over $\mathbf{S}$. Examples of such problems include regulation (tracking and disturbance rejection) and decoupling.

Consider the linear equation $AR = B$, where A and B are given matrices of compatible dimensions and R is unknown. If A, B, R are matrices over a field, then the following result is well-known: There exists an R such that $AR = B$ if and only if rank $A = $ rank $[A \ B]$. However, if A, B, R are matrices over a ring, the rank condition alone is not enough.

Lemma 5.7.1   *Suppose $\mathbf{R}$ is a commutative ring with identity, and that $A, B \in \mathbf{M(R)}$ have the same number of rows. Then there exists an $R \in \mathbf{M(R)}$ such that $AR = B$ if and only if $[A \ 0]$ and $[A \ B]$ are right associates.*

Proof. "if" Suppose $U \in \mathbf{U(R)}$ is such that $[A \ B] = [A \ 0]U$. Partitioning U as a block 2×2 matrix in the obvious way leads to

$$[A \ B] = [A \ 0] \begin{bmatrix} U_{11} & U_{12} \\ U_{21} & U_{22} \end{bmatrix}, \tag{5.7.1}$$

or $B = AU_{12}$.

"only if" Suppose $B = AR$ for some $R \in \mathbf{M(R)}$. Then

$$[A \ B] = A[I \ R] =_R A[I \ R] \begin{bmatrix} I & -R \\ 0 & I \end{bmatrix}$$

$$= A[I \ 0] = [A \ 0], \tag{5.7.2}$$

where $=_R$ denotes right equivalence (see Section B.1). Hence, $[A \ B]$ and $[A \ 0]$ are right associates. $\square$

The next lemma presents another set of conditions for the equation $AR = B$ to have a solution. It is a bit more explicit than Lemma 5.7.1. However, it requires that the ring $\mathbf{R}$ be a principal ideal domain.

Lemma 5.7.2   *Suppose $\mathbf{R}$ is a p.i.d., and that $A, B \in \mathbf{M(R)}$ have the same number of rows. Let r denote the rank of A, and let $a_1, \cdots, a_r$ denote the invariant factors of A. Suppose $U, V \in \mathbf{U(R)}$ are unimodular* *matrices such that* $U A V = S$, *where*

$$
S = \begin{bmatrix}
a_1 & & & & \\
 & \cdot & & & 0 \\
 & & \cdot & & \\
 & & & \cdot & \\
 & & & & a_r \\
 & & 0 & & 0
\end{bmatrix},
\tag{5.7.3}
$$

is the Smith form of A. Then there exists an $R \in \mathbf{M}(\mathbf{R})$ such that $A R = B$ if and only if
 (i) For $i = 1, \cdots , r$, a_i divides each element in the i-th row of $U B$, and
 (ii) All other rows of $U B$ (if any) are zero.

Proof. Note that $A = U^{-1} S V^{-1}$, so that the equation $A R = B$ is equivalent to $S V^{-1} R = U B$. Since V is unimodular, we can replace $V^{-1} R$ by another unknown matrix T, so that the equation now becomes $S T = U B$. From this (i) and (ii) follow readily. $\square$

To set up the regulation and decoupling problems, consider the configuration shown in Figure 5.8, where P is a given plant, and T and R are specified reference signal generators. To achieve the greatest possible generality, a two-parameter compensation scheme is shown in this figure. The objective is to find a $C \in S(P)$ that achieves one of the following additional features:
 (i) *Tracking*: $(I - H_{yt})T \in \mathbf{M}(\mathbf{S})$.
 (ii) *Disturbance Rejection*; $H_{yd} R \in \mathbf{M}(\mathbf{S})$.
 (iii) *Decoupling*: H_{yt} is block-diagonal (and nonzero).

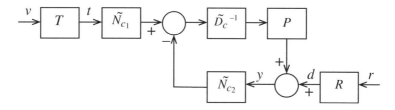

Figure 5.8: Configuration for Tracking and Disturbance Rejection.

Thus, tracking corresponds to the transfer matrix from v to the tracking error $t - y$ being stable; disturbance rejection corresponds to the transfer matrix from r to the plant output y being stable. It is clear that, if the compensation scheme stabilizes the plant, then tracking and disturbance rejection pose additional restrictions only when T and R are unstable. Decoupling corresponds to the transfer matrix from the external input (also taken to be t) to the plant output being diagonal, or more generally block diagonal.

The following result summarizes a simplification that is used repeatedly.

Lemma 5.7.3   *Suppose P, Q, $R \in \mathbf{M}(\mathbf{S})$ with $|Q| \neq 0$ and Q, R left-coprime. Then $PQ^{-1}R \in \mathbf{M}(\mathbf{S})$ if and only if $PQ^{-1} \in \mathbf{M}(\mathbf{S})$.*

Proof. "if" Obvious.

"only if" Select A, $B \in \mathbf{M}(\mathbf{S})$ such that $QA + RB = I$. Then

$$Q^{-1} = A + Q^{-1}RB\,, \tag{5.7.4}$$
$$PQ^{-1} = PA + PQ^{-1}RB\,. \tag{5.7.5}$$

Hence, if $PQ^{-1}R \in \mathbf{M}(\mathbf{S})$, then $PQ^{-1} \in \mathbf{M}(\mathbf{S})$. $\square$

We are now in a position to state the main results of this section.

Theorem 5.7.4   Consider the system of Figure 5.8, and let $(\tilde{D}_t, \tilde{N}_t)$, $(\tilde{D}_r, \tilde{N}_r)$ be any l.c.f.'s of T and R, respectively. Let (N, D), $(\tilde{D}, \tilde{N})$ be any r.c.f. and any l.c.f. of P, and let $X, Y \in \mathbf{M}(\mathbf{S})$ be any particular solutions of the identity $XN + YD = I$. Then:

(i) The tracking problem has a solution if and only if there exist $Q, W \in \mathbf{M}(\mathbf{S})$ such that

$$NQ + W\tilde{D}_t = I\,. \tag{5.7.6}$$

(ii) The disturbance rejection problem has a solution if and only if there exist $S, V \in \mathbf{M}(\mathbf{S})$ such that

$$NS\tilde{D} + W\tilde{D}_r = NX\,. \tag{5.7.7}$$

Proof. By Theorem 5.6.3, the set of stable transfer matrices from (t, r) to y is given by

$$[NQ \quad -N(X + S\tilde{D})]\,, \quad Q, S \in \mathbf{M}(\mathbf{S})\,. \tag{5.7.8}$$

Thus, the tracking problem has a solution if and only if there exists a matrix $Q \in \mathbf{M}(\mathbf{S})$ such that

$$(I - NQ)\tilde{D}_t^{-1}\tilde{N}_t \in \mathbf{M}(\mathbf{S})\,, \tag{5.7.9}$$

But by Lemma 5.7.3, (5.7.9) is equivalent to

$$(I - NQ)\tilde{D}_t^{-1} \in \mathbf{M}(\mathbf{S})\,, \tag{5.7.10}$$

for some $Q \in \mathbf{M}(\mathbf{S})$. Now (5.7.10) is the same as

$$I - NQ = W\tilde{D}_t\,, \tag{5.7.11}$$

for some $Q, W \in \mathbf{M}(\mathbf{S})$. It is clear that (5.7.11) and (5.7.6) are equivalent. One can derive (5.7.7) in the same manner. $\square$

The criteria given in Theorem 5.7.4 are testable, since (5.7.6) and (5.7.7) are just linear equations in the components of the unknown matrices Q, W, S, V, so that their solvability can be ascertained using Lemma 5.7.1 and 5.7.2.

It is left to the reader to derive the condition analogous to (5.7.6) in the case where a one-parameter compensator is used.

Finally, consider the problem of decoupling. It is easy to see that if N is square, then NQ can always be made into a diagonal matrix by an appropriate choice of Q (e.g., $Q = N^{adj}$). More generally, if N has more columns than rows, then once again it is possible to make NQ into a diagonal matrix.

NOTES AND REFERENCES

The formulation of the feedback stabilization problem is from [19], which also contains several examples that show that any three of the components of $H(P, C)$ could be stable while the fourth is unstable.

The parametrization of all compensators that stabilize a given plant is given in [111]. This characterization is somewhat different from the one given here, in that it involves a combination of both stable rational functions as well as polynomials. The present treatment is due to [25], with some technical improvements from [101].

The notion of strong stabilization is defined, and necessary and sufficient conditions for it are derived, in [108]. The current approach based on matrix Euclidean division is from [102]. See [2] for an alternate set of conditions for strong stabilizability based on Cauchy indices; these conditions do not require one to compute explicitly the poles and zeros of a plant.

The present results on simultaneous stabilization are from [102]; see [83] for a geometric interpretation of this problem. The interlacing test for simultaneous stabilization is from [66, 84, 85]. The idea of reliable stabilization using a multi-compensator configuration is introduced in [87, 88], while the present results are taken from [103]. The treatment of two-parameter compensators follows [22].

APPENDIX A

Algebraic Preliminaries

This appendix contains a brief introduction to some basic properties of rings. It is important to note that the choice of results from ring theory that are presented here is highly selective; the reader is referred to [52] or [116] for a more comprehensive treatment.

A.1 RINGS, FIELDS AND IDEALS

Definition A.1.1 A *ring* is a nonempty set $\mathbf{R}$ together with two binary operations $+$ (addition) and $\cdot$ (multiplication) such that the following axioms are satisfied:

(R1) $(\mathbf{R}, +)$ is a commutative group. This means that

$a + (b + c) = (a + b) + c, \forall a, b, c \in \mathbf{R}$.

$a + b = b + a, \forall a, b \in \mathbf{R}$.

There exists an element $0 \in \mathbf{R}$ such that

$a + 0 = 0 + a = a, \forall a \in \mathbf{R}$.

For every element $a \in \mathbf{R}$, there exists a corresponding element $-a \in \mathbf{R}$ such that $a + (-a) = 0$.

(R2) $(\mathbf{R}, \cdot)$ is a semigroup. This means that

$a \cdot (b \cdot c) = (a \cdot b) \cdot c, \forall a, b, c \in \mathbf{R}$.

(R3) Multiplication is distributive over addition. This means that

$a \cdot (b + c) = a \cdot b + a \cdot c$,

$(a + b) \cdot c = a \cdot c + b \cdot c, \forall a, b, c \in \mathbf{R}$.

As is customary, $a \cdot b$ is denoted by ab, and $a + (-b)$ is denoted by $a - b$.

A ring $\mathbf{R}$ is said to be *commutative* if $ab = ba \, \forall a, b \in \mathbf{R}$, and is said to *have an identity* if there exists an element $1 \in \mathbf{R}$ such that $1 \cdot a = a \cdot 1 = a \, \forall a \in \mathbf{R}$.

Examples A.1.2 A classic example of a ring is the set of integers, denoted by $\mathbf{Z}$, with the usual notions of addition and multiplication. Clearly, $\mathbf{Z}$ is a commutative ring with identity. The set $\mathbf{E}$ of even integers, together with the usual addition and multiplication is an example of a ring without identity; but **E** is commutative. The set $\mathbf{Z}^{2\times2}$ of 2×2 matrices with integer elements is a noncommutative ring with identity.

A ring **R** is said to be a *domain* (or an integral domain) if $a \in \mathbf{R}$, $b \in \mathbf{R}$, $ab = 0$ implies that either $a = 0$ or $b = 0$. In other words, **R** is a domain if the product of every pair of nonzero elements is nonzero.

Example A.1.3   In Example A.1.2 above, both **Z** and **E** are domains, but $\mathbf{Z}^{2\times2}$ is not.

Example A.1.4   Consider the ring $\mathbf{Z}/(6)$ consisting of the six elements $\{0, 1, 2, 3, 4, 5\}$, with addition and multiplication defined modulo 6. For example, $3 + 5 = 8 \mod 6 = 2$, and $2.5 = 10 \mod 6 = 4$. This ring is commutative and has an identity. However, since $2 \cdot 3 = 0$, it is not a domain.

Suppose **R** is a ring with identity. An element $x \in \mathbf{R}$ is called a *unit* of **R** if there is a $y \in \mathbf{R}$ such that $xy = yx = 1$. It can be easily shown that such a y is unique; y is called the *inverse* of x and is denoted by x^{-1}.

Example A.1.5   Let $\mathbf{R} = \mathbf{C}[0, 1]$, the set of continuous real-valued functions defined over the interval $[0, 1]$. If addition and multiplication on **R** are defined pointwise, i.e.,

$$(x + y)(t) = x(t) + y(t) ,$$
$$(xy)(t) = x(t)y(t), \forall t \in [0, 1], \forall x, y \in \mathbf{R} ,$$

then **R** is a commutative ring with identity. However, it is not a domain; for example, let

$$x(t) = \begin{cases} 1 - 2t & 0 \le t \le 0.5 \\ 0 & 0.5 < t \le 1 \end{cases}$$
$$y(t) = \begin{cases} 0 & 0 \le t \le 0.5 \\ 2t - 1 & 0.5 < t \le 1 \end{cases}.$$

Then $xy = 0$, even though $x \neq 0$, $y \neq 0$. A function $x \in \mathbf{R}$ is a unit of **R** if and only if $x(t) \neq 0 \, \forall t \in [0, 1]$. In other words, the units in **R** are the functions that do not change sign over $[0, 1]$.

Definition A.1.6   A *field* is a commutative ring **F** with an identity, satisfying two additional assumptions:
   (F1) **F** contains at least two elements.
   (F2) Every nonzero element of **F** is a unit.

Examples A.1.7   The rational numbers, real numbers and complex numbers are all well-known examples of fields. Another example is $\mathbf{Z}/(p)$ where p is a prime integer, consisting of the elements

$\{0, 1, \cdots, p - 1\}$. Addition and multiplication are defined modulo p. This set is clearly a commutative ring with identity, and it contains at least two elements. To show that it satisfies axiom (F2), suppose a is any nonzero element of $\mathbf{Z}/(p)$. Since the greatest common divisor of a and p (in the usual sense) is 1, there exist integers x and y such that

$$ax + py = 1 .$$

Moreover, for *any* integer q, we have that

$$a(x - qp) + p(y + qa) = 1 .$$

So by a suitable choice of q, the integer $x - qp$ can be made to lie between 0 and $p - 1$. As a result, it follows that there exist integers b and c, with b lying between 0 and $p - 1$, such that

$$ab + pc = 1, \text{ or } ab = 1 - pc \equiv 1 \mod p .$$

Hence, a is a unit and b is its inverse. For example, in the ring $\mathbf{Z}/(13)$, the inverse of 5 is 8, and the inverse of 2 is 7.

The field $\mathbf{Z}/(2)$, consisting of the two elements $\{0, 1\}$, is called the *binary field*.

A subset $\mathbf{S}$ of a ring $\mathbf{R}$ is a *subring* of $\mathbf{R}$ if it is a ring in its own right, i.e., if $0 \in \mathbf{S}$, and the sum, difference and product of two elements of $\mathbf{S}$ again belong to $\mathbf{S}$.

Definition A.1.8   A subset $\mathbf{I}$ in a ring $\mathbf{R}$ is said to be a *left ideal* if (i) $\mathbf{I}$ is a subgroup of the additive group of $\mathbf{R}$, and (iiL) $a \in \mathbf{I}$, $x \in \mathbf{R}$ imply that $xa \in \mathbf{I}$. $\mathbf{I}$ is a *right ideal* if (i) $\mathbf{I}$ is a subgroup of the additive group of $\mathbf{R}$ and (iiR) $a \in \mathbf{I}$, $x \in \mathbf{R}$ imply that $ax \in \mathbf{I}$. $\mathbf{I}$ is an *ideal* if it is both a left ideal and a right ideal.

In the above definition, condition (i) means the following: First, $0 \in \mathbf{I}$, and second, if $x, y \in \mathbf{I}$, then $x \pm y \in \mathbf{I}$. Condition (iiL) (resp. (iiR)) means that if and element of $\mathbf{I}$ is multiplied on the left (resp. right) by any element of $\mathbf{R}$, the product is once again in $\mathbf{I}$. Clearly, if $\mathbf{R}$ is a commutative ring, the concepts of a left ideal, right ideal, and ideal all coincide.

Example A.1.9   Let $\mathbf{R}$ be the ring $\mathbf{Z}^{2 \times 2}$ of 2×2 matrices with integer elements. Let

$$\mathbf{I}_1 = \{M \in \mathbf{Z}^{2 \times 2} : m_{11} = 0, m_{21} = 0\} .$$

Then it is easy to verify that $\mathbf{I}_1$ is a left ideal in $\mathbf{R}$. Similarly,

$$\mathbf{I}^2 = \{M \in \mathbf{Z}^{2 \times 2} : m_{21} = 0, m_{22} = 0\}$$

is a right ideal in $\mathbf{R}$. The set of diagonal matrices

$$\mathbf{D} = \{M \in \mathbf{Z}^{2 \times 2} : m_{21} = 0, m_{12} = 0\}$$

is a subring of **R** but is neither a left ideal nor a right ideal.

Let a be an integer. Then

$$\mathbf{M}(a) = \{M \in \mathbf{Z}^{2 \times 2} : m_{ij} \text{ is divisible by } a \ \forall i, j \}$$

is an ideal in **R**.

Example A.1.10   Consider the ring $C[0, 1]$ of Example A.1.5, and let $t_0 \in [0, 1]$. Then

$$\mathbf{I}_{t_0} = \{x(\cdot) \in C[0, 1] : x(t_0) = 0\}$$

is an ideal. More generally, let **S** be any subset of $[0, 1]$. Then

$$\mathbf{I}_S = \{x(\cdot) \in C[0, 1] : x(t) = 0 \ \forall t \in S\}$$

is an ideal.

Suppose a is some element of a ring **R**. Then the set of all elements of the form xa where $x \in \mathbf{R}$, i.e., the set of all left multiples of a, is a left ideal. It is called the *left principal ideal* generated by a. Similarly, the *right principal ideal* generated by a is the set of all elements ax where $x \in \mathbf{R}$.

An ideal **I** in a commutative ring **R** is a *prime ideal* if $a \in \mathbf{R}, b \in \mathbf{R}, ab \in \mathbf{I}$ implies that either $a \in \mathbf{I}$ or $b \in \mathbf{I}$. Equivalently, **I** is a prime ideal if and only if the set $\mathbf{R} - \mathbf{I}$ is closed under multiplication (i.e., the product of two elements not belonging to **I** cannot itself belong to **I**).

Example A.1.11   Consider the ring of integers **Z**, and let n be any integer. Then the set of multiples of n, denoted by (n), is the principal ideal generated by n. It is a prime ideal if and only if n divides ab implies that n divides a or n divides b, which is true if and only if n is a prime number.

Example A.1.12   Consider the ideal $\mathbf{I}_{t_0}$ of Example A.1.10. If a product xy belongs to this ideal, then $(xy)(t_0) = 0$, which means that either $x(t_0) = 0$ or $y(t_0) = 0$. Hence, $\mathbf{I}_{t_0}$ is a prime ideal. However, if **S** contains more than one point, then $\mathbf{I}_S$ is not a prime ideal.

By convention, the entire ring **R** is *not* considered to be a prime ideal. This is in conformity with the convention that 1 is not considered a prime number (see Example A.1.11). Thus, "**I** is a prime ideal in **R**" automatically implies, by convention, that **I** is a *proper* subset of **R**.

PROBLEMS

A.1.1. (i) Show that the zero element of a ring is unique (Hint: if 0_1 and 0_2 are two additive identities, then $0_1 = 0_1 + 0_2 = 0_2$).

(ii) Show that the additive inverse of an element is unique.

A.1.2. (i) Show that $a \cdot 0 = 0$ for all a in a ring (Hint: $ab = a(b + 0)$).

(ii) Show that, if a ring $\mathbf{R}$ contains an identity and has at least one nonzero element, then $1 \neq 0$ (Hint: let $a \neq 0$; then $a0 \neq a = a1$).

A.1.3. Show that, if $\mathbf{R}$ is a domain, then the cancellation law holds, i.e., $ab = ac, a \neq 0$ implies that $b = c$.

A.1.4. Let $\mathbf{R}$ be a ring with identity and at least one nonzero element.

(i) Show that the set of units of $\mathbf{R}$ is a group under multiplication; i.e., show that (a) if u is a unit, then so is u^{-1}, and (b) if u, v are units, so are uv and vu.

(ii) Show that 0 can never be a unit (Hint: see Problem A.1.2).

(iii) Show that if x has an inverse, then it is unique.

(iv) Show that if x has a left inverse y and a right inverse z, then $y = z$ (i.e., show that $yx = 1, xz = 1$ implies $y = z$).

A.1.5. Let $\mathbf{R}$ be the set of functions mapping the interval $[0, 1]$ into the set of integers $\mathbf{Z}$.

(i) Show that $\mathbf{R}$ is a ring under pointwise addition and multiplication.

(ii) What are the units of this ring?

A.1.6. Consider the ring $\mathbf{Z}/(p)$ of Example A.1.7. Show that this ring is a field if and only if p is a prime number.

A.1.7. (i) Consider the ring $\mathbf{Z}/(9)$, consisting of $\{0, 1, \cdots, 8\}$, with addition and multiplication modulo 9. Determine the units of this ring. (Answer: 1,2,4,5,7,8)

(ii) Consider the ring $\mathbf{Z}/(n)$. Show that m is a unit of this ring if and only if the greatest common divisor of m and n is 1.

A.1.8. Consider the ring $\mathbf{Z}^{n \times n}$ of $n \times n$ matrices with integer elements. Show that a matrix M in this ring is a unit if and only if its determinant is ± 1.

A.1.9. Show that a commutative ring $\mathbf{R}$ is a domain if and only if $\{0\}$ is a prime ideal.

A.2 RINGS AND FIELDS OF FRACTIONS

Throughout this section "ring" means a commutative ring with identity.

Suppose $\mathbf{R}$ is a ring. An element $a \in \mathbf{R}$ is an *absolute nondivisor of zero* if $b \in \mathbf{R}$, $ab = 0$ implies that $b = 0$. In every ring, there are absolute nondivisors of zero; for example, all the units are of this type. But there may be others as well.

Examples A.2.1   In the ring $\mathbf{Z}^{n \times n}$, a matrix M is an absolute nondivisor of zero if and only if the determinant of M is nonzero. Thus, a nonunit matrix can be an absolute nondivisor of zero (cf. Problem A.1.8).

Now consider the ring $C[0, 1]$ defined in Example A.1.5. Suppose a function $x(\cdot)$ belonging to this ring vanishes at only a finite number of points. If $y \in C[0, 1]$ and $xy = 0$, then $x(t)y(t) \equiv 0$, which means that $y(t) = 0$ for all except a finite number of values of t. However, since $y(\cdot)$ is continuous, this implies that $y(t) \equiv 0$, or that y is the zero element of the ring. Hence, x is an absolute nondivisor of zero.

If a is an absolute nondivisor of zero and $ab = ac$, then $b = c$; in other words, absolute nondivisors of zero can be "cancelled."

Of course, if $\mathbf{R}$ is a domain, then *every* nonzero element of $\mathbf{R}$ is an absolute nondivisor of zero.

A set $\mathbf{M}$ in a ring $\mathbf{R}$ is said to be a *multiplicative system* if $a, b \in \mathbf{M}$ implies that $ab \in \mathbf{M}$. It is *saturated* if $a \in \mathbf{R}$, $b \in \mathbf{R}$, $ab \in \mathbf{M}$ implies that $a \in \mathbf{M}$, $b \in \mathbf{M}$.

Fact A.2.2   The set $\mathbf{N}$ of absolute nondivisors of zero in a ring $\mathbf{R}$ is a multiplicative system.

Proof. Suppose $a, b \in \mathbf{N}$, $y \in \mathbf{R}$, and $aby = 0$. Then, since $a, b \in \mathbf{N}$, it follows successively that $aby = 0 \Longrightarrow by = 0 \Longrightarrow y = 0$. Hence, $ab \in \mathbf{N}$. $\square$

Suppose $\mathbf{R}$ is a ring, $\mathbf{M}$ is a multiplicative system in $\mathbf{R}$ containing 1, and $\mathbf{M}$ is a subset of $\mathbf{N}$ (the set of absolute nondivisors of zero in $\mathbf{R}$). We now begin a construction which will ultimately result in a ring $\mathbf{L}$ which contains $\mathbf{R}$ as a subring, and in which every element of $\mathbf{M}$ is a unit.

Consider the set $\mathbf{R} \times \mathbf{M}$, and define a binary relation $\sim$ on $\mathbf{R} \times \mathbf{M}$ as follows: $(a, b) \sim (c, d) \Longleftrightarrow ad = bc$. The relation $\sim$ is an equivalence relation: Clearly $\sim$ is reflexive and symmetric. To show that it is transitive, suppose $(a, b) \sim (c, d)$ and $(c, d) \sim (e, f)$. Then $ad = bc$ and $cf = de$. Multiplying the first equation by f and the second one by b gives $adf = bcf = bde$. Now $d \in \mathbf{N}$ since $d \in \mathbf{M}$ and $\mathbf{M} \subseteq \mathbf{N}$. Hence, d can be cancelled in the above equation to give $af = be$, i.e., $(a, b) \sim (e, f)$.

Since $\sim$ is an equivalence relation, the set $\mathbf{R} \times \mathbf{M}$ can be partitioned into disjoint equivalence classes under $\sim$. The set of equivalence classes $\mathbf{R} \times \mathbf{M}/ \sim$ is denoted by $\mathbf{L}$. The set $\mathbf{L}$ therefore consists of *fractions* (a, b), or a/b in more familiar terms, where we agree to treat two fractions a/b

and c/d as equivalent if $ad = bc$. Addition and multiplication of fractions in $\mathbf{L}$ are defined in the familiar way, namely

$$\frac{a}{b} + \frac{c}{d} = \frac{ad + bc}{bd} . \tag{A.1}$$

$$\frac{a}{b} \cdot \frac{c}{d} = \frac{ac}{bd} . \tag{A.2}$$

Note that, if $b, d \in \mathbf{M}$, then so does bd. Hence, the right sides of (A.1) and (A.2) are valid fractions. Actually, (A.1) and (A.2) represent operations on equivalence classes; but the reader can verify that the sum and product of two fractions (i.e., two equivalence classes) do not depend on which representatives of the equivalence classes are used.

With addition and multiplication defined by (A.1) and (A.2), $\mathbf{L}$ is a ring. Moreover, if every $a \in \mathbf{R}$ is identified with the fraction $a/1$, then $\mathbf{R}$ is isomorphic to a subring of $\mathbf{L}$. The element $1/1$ serves as an identity for $\mathbf{L}$. Finally, if $d \in \mathbf{M}$, then d corresponds to $d/1 \in \mathbf{L}$; moreover, $d/1$ is a unit in $\mathbf{L}$ with the inverse $1/d$. Thus, every element of (the isomorphic image of) $\mathbf{M}$ is a unit in $\mathbf{L}$. The ring $\mathbf{L}$ is called the *ring of fractions* of $\mathbf{R}$ with respect to $\mathbf{M}$, and is denoted by $\mathbf{M}^{-1}\mathbf{R}$.

Since $\mathbf{M}$ is a subset of $\mathbf{N}$ (the set of absolute nondivisors of zero), the ring $\mathbf{N}^{-1}\mathbf{R}$ contains the largest number of units. If in particular $\mathbf{R}$ is a domain, then $\mathbf{N} = \mathbf{R} \setminus 0$, i.e., $\mathbf{N}$ consists of all nonzero elements of $\mathbf{R}$. In this case, the ring $\mathbf{F} = (\mathbf{R} \setminus 0)^{-1}\mathbf{R}$ has the property that *every* nonzero element of $\mathbf{R}$ is a unit in $\mathbf{F}$. Moreover, every nonzero element of $\mathbf{F}$ is also a unit: If $a/b \in \mathbf{F}$ and $a \neq 0$, then b/a is the inverse of a/b. Hence, $\mathbf{F}$ is a field; it is referred to as the *field of fractions* or *quotient field* associated with the domain $\mathbf{R}$.

There is a particularly important class of fraction rings that is frequently encountered. Suppose $\mathbf{R}$ is a ring, and that $\mathbf{I}$ is a prime ideal in $\mathbf{R}$. If $\mathbf{M}$ denotes the complement of $\mathbf{I}$, then $\mathbf{M}$ is a multiplicative system (see Problem A.1.4). The corresponding fraction ring $\mathbf{M}^{-1}\mathbf{R}$ is called the *localization* of $\mathbf{R}$ with respect to $\mathbf{I}$.

Examples A.2.3   The set of integers $\mathbf{Z}$ is a domain. The field of fractions associated with the integers is the set of rational numbers.

In the ring $\mathbf{Z}$, let $\mathbf{M}$ denote the set of numbers that are *not* divisible by 3. Then $\mathbf{M}$ is a multiplicative system (since 3 is a prime number), and $1 \in \mathbf{M}$. The ring of fractions $\mathbf{M}^{-1}\mathbf{Z}$ is the set of rational numbers whose denominators (when expressed in reduced form) are not divisible by 3. This is a *subring* of the *field* of rational numbers.

PROBLEMS

A.2.1. Suppose $\mathbf{R}$ is a ring, $a, b, c \in \mathbf{R}$, and a is an absolute nondivisor of zero. Show that if $ab = ac$, then $b = c$.

A.2.2. Show that the addition and multiplication rules (A.1) and (A.2) are unambiguous, in the sense that the final answers do not depend on which representatives of the equivalence classes are used in the computation.

A.2.3. Consider the ring $\mathbf{Z}/(10)$, consisting of the integers $0, \cdots , 9$ with addition and multiplication defined modulo 10.

(i) Determine the units of this ring.

(ii) Show that the set $\mathbf{N}$ of absolute nondivisors of zero consists of just the units.

(iii) Using (ii), show that the ring of fractions with numerators in $\mathbf{Z}/(10)$ and denominators in $\mathbf{N}$ is again just $\mathbf{Z}/(10)$.

A.2.4. Suppose $\mathbf{R}$ is a ring and that $\mathbf{I}$ is a prime ideal in $\mathbf{R}$. Let $\mathbf{M}$ denote the complement of $\mathbf{I}$. Show that $\mathbf{M}$ is a multiplicative system.

A.3   PRINCIPAL IDEAL DOMAINS

Throughout this section, "ring" means a commutative ring with identity.

Definition A.3.1   A ring $\mathbf{R}$ is said to be a *principal ideal ring* if every ideal in $\mathbf{R}$ is principal. $\mathbf{R}$ is a *principal ideal domain (p.i.d.)* if it is a domain as well as a principal ideal ring.

Recall that a principal ideal $\mathbf{I}$ consists of *all* multiples of some element a, i.e., $\mathbf{I} = \{xa : x \in \mathbf{R}\}$. Thus, in a principal ideal ring, every ideal is generated by a single element.

If x and y are elements of a ring $\mathbf{R}$ with $x \neq 0$, we say that x *divides* y, and y *is a multiple* of x, if there is an element $z \in \mathbf{R}$ such that $y = xz$; this is denoted by $x \mid y$. If x and y are elements of a ring $\mathbf{R}$ such that not both are zero, a *greatest common divisor (GCD)* of x, y is any element $d \in \mathbf{R}$ such that

$$(\text{GCD1}) \ d \mid x \text{ and } d \mid y \,.$$
$$(\text{GCD2}) \ c \mid x, c \mid y \Longrightarrow c \mid d \,.$$

In the above definition, it is implicit that $d \neq 0$, since d divides a nonzero element by virtue of (GCD1). We say *a* greatest common divisor because a GCD is not unique. Certainly, if d is a GCD of x and y, then so is $-d$. Actually, the following stronger result holds:

Fact A.3.2   Suppose d is a GCD of x and y; then so is du whenever u is a unit. Suppose in addition that $\mathbf{R}$ is a domain; then every GCD d_1 of x and y is of the form $d_1 = du$ for some unit u.

Proof. The first sentence is obvious. To prove the second sentence, observe that if d_1 is another GCD of x and y, then by (GCD2) $d_1 \mid d$ and $d \mid d_1$. Hence, $d_1 = du$ for some u. Moreover, $du \mid d$ since $d_1 \mid d$. Since $\mathbf{R}$ is a domain and $d \neq 0$, this implies that $u \mid 1$, i.e., that u is a unit. $\square$

Fact A.3.2 states that if $\mathbf{R}$ is a domain, then once we have found *one* GCD of a given pair of elements, we can quickly find them *all*. However, a question that is not answered by Fact A.3.2 is: Does every pair of elements have a GCD? The answer to this question is provided next.

Theorem A.3.3   Suppose $\mathbf{R}$ is a principal ideal ring. Then every pair of elements $x, y \in \mathbf{R}$, not both of which are zero, has a GCD d which can be expressed in the form

$$d = px + qy \tag{A.1}$$

for appropriate elements $p, q \in \mathbf{R}$. Moreover, if $\mathbf{R}$ is a domain, then *every* GCD of x and y can be expressed in the form (A.1).

Proof. Given $x, y \in \mathbf{R}$, consider the set

$$\mathbf{I} = \{ax + by, a, b \in \mathbf{R}\} . \tag{A.2}$$

In other words, $\mathbf{I}$ is the set of all "linear combinations" of x and y. It is easy to verify that $\mathbf{I}$ is an ideal in $\mathbf{R}$.[1] Since $\mathbf{R}$ is a principal ideal ring, $\mathbf{I}$ must be a principal ideal; that is, there exists an element $d \in \mathbf{I}$ such that $\mathbf{I}$ is the set of all multiples of d. Since $x = 1 \cdot x + 0 \cdot y \in \mathbf{I}$, it follows that x is a multiple of d, i.e., $d \mid x$; similarly $d \mid y$. Thus, d satisfies the axiom (GCD1). Next, since $d \in \mathbf{I}$ and $\mathbf{I}$ is the ideal generated by x, y, there exist $p, q \in \mathbf{R}$ such that (A.1) holds. Now suppose $c \mid x, c \mid y$. Then $c \mid (px + qy)$, i.e., $c \mid d$. Hence, d also satisfies (GCD2) and is thus a GCD of x and y.

Up to now, we have shown that every pair of elements x and y has *a* GCD which can be written in the form (A.1) for a suitable choice of p and q. Now, if in addition $\mathbf{R}$ is a domain, then *every* GCD d_1 of x, y is of the form $d_1 = du$ for some unit u, by Fact A.3.2. Thus, every GCD d_1 can be written in the form $d_1 = du = upx + upy$.   □

Two elements $x, y \in \mathbf{R}$ are *relatively prime* or simply *coprime* if every GCD of x, y is a unit. In view of Fact A.3.2, if $\mathbf{R}$ is a domain, this is equivalent to saying that x and y are coprime if and only if 1 is a GCD of x and y. Now suppose $\mathbf{R}$ is a principal ideal domain. Then, by Theorem A.3.3, $x, y \in \mathbf{R}$ are coprime if and only if the ideal $\mathbf{I}$ in (A.2) is the same as the ideal generated by 1. But the latter is clearly the entire ring $\mathbf{R}$. This can be summarized as follows:

Fact A.3.4   Let $\mathbf{R}$ be a principal ideal domain. Then $x, y \in \mathbf{R}$ are coprime if and only if there exist $p, q \in \mathbf{R}$ such that $px + qy = 1$.

One can also define a GCD of an n-tuple of elements $(x_1, \cdots, x_n)$, not all of which are zero. An element $d \in \mathbf{R}$ is a GCD of the collection $(x_1, \cdots, x_n)$ if

$$\begin{array}{ll} (i) & d \mid x_i \; \forall i , \hfill \text{(A.3)} \\ (ii) & c \mid x_i \; \forall i \implies c \mid d . \hfill \text{(A.4)} \end{array}$$

---

[1] $\mathbf{I}$ is referred to as the *ideal generated by x and y*.

As in Fact A.3.2, it follows that a GCD of a given n-tuple is unique to within a unit provided the ring is a domain. Moreover, Theorem A.3.3 can be generalized as follows:

Theorem A.3.5   Suppose $\mathbf{R}$ is a principal ideal ring, and suppose $x_1, \cdots, x_n \in \mathbf{R}$, with at least one element not equal to zero. Then the n-tuple $(x_1, \cdots, x_n)$ has a GCD d which can be expressed in the form

$$d = \sum_{i=1}^{n} p_i x_i \tag{A.5}$$

for appropriate elements $p_1, \cdots, p_n \in \mathbf{R}$. Moreover, if $\mathbf{R}$ is a domain, then every GCD of this n-tuple can be expressed in the form (A.5).

The proof is entirely analogous to that of Theorem A.3.3 and is left as an exercise.

In the field of rational numbers, every ratio a/b of integers has an equivalent "reduced form" f/g where f and g are coprime. The next result shows that such a statement is also true in the field of fractions associated with *any* principal ideal domain.

Fact A.3.6   Let $\mathbf{R}$ be a principal ideal domain, and let $\mathbf{F}$ be the field of fractions associated with $\mathbf{R}$. Given any fraction a/b in $\mathbf{F}$, there exists an equivalent fraction f/g such that f and g are coprime.

Proof. If a and b are already coprime, then there is nothing to be done. Otherwise, let d be a GCD of a, b, and define $f = a/d, g = b/d$. Then clearly, $f/g = a/b$, and it only remains to show that f and g are coprime. From Theorem A.3.3, there exist $p, q \in \mathbf{R}$ such that

$$d = pa + qb = pfd + qgd . \tag{A.6}$$

Cancelling d from both sides of (A.6) gives $1 = pf + qg$, which shows that f, g are coprime.   □

Thus, the procedure for "reducing" a fraction is the natural one: namely, we extract a greatest common divisor from the numerator and the denominator.

Two elements $a, b \in \mathbf{R}$ are *associates* (denoted by $a \sim b$) if there is a unit u such that $a = bu$. One can readily verify that $\sim$ is an equivalence relation on $\mathbf{R}$. A nonunit, nonzero element $p \in \mathbf{R}$ is a *prime* if the only divisors of p are either units or associates of p. An equivalent definition is the following: p is a prime if $p = ab, a, b \in \mathbf{R}$ implies that either a or b is a unit. In the ring of integers, the primes are precisely the prime numbers.

A useful property of principal ideal domains is stated next, without proof (see [116]).

Fact A.3.7   Every nonunit, nonzero element of a principal ideal domain can be expressed as a product of primes. Moreover, this factorization is unique in the following sense: If

$$x = \prod_{i=1}^{n} p_i = \prod_{i=1}^{m} q_i , \tag{A.7}$$

where p_i, q_i are all primes, then $n = m$, and the q_i's can be renumbered such that $p_i \sim q_i \; \forall i$.

Thus, the only nonuniqueness in the prime factorization of an element arises from the possibility of multiplying some of the prime factors by a unit (for example, $6 = 2 \cdot 3 = (-2) \cdot (-3)$).

Using prime factorizations, one can give a simple expression for a GCD of a set of elements. The proof is left as an exercise (see Problem A.3.7).

Fact A.3.8   Suppose $x_1, \cdots, x_n$ is a set of elements such that none is a unit nor zero. Express each element x_i in terms of its prime factors in the form

$$x_i = \prod_{i=1}^{n} p_j^{\alpha_{ij}} \tag{A.8}$$

where the p_j's are distinct (i.e., nonassociative) primes, $\alpha_{ij} \geq 0$ and $\alpha_{ij} = 0$ if p_j is not a divisor of x_i. Then a GCD d of $(x_1. \cdots, x_n)$ is given by

$$d = \prod_{i=1}^{n} p_j^{\beta_j} \tag{A.9}$$

where

$$\beta_j = \min_i \alpha_{ij} . \tag{A.10}$$

Corollary A.3.9   *Two elements x, y are coprime if and only if their prime divisors are distinct.*

Up to now, we have talked about the greatest common divisor of a set of elements. A parallel concept is that of the least common multiple. Suppose $(x_1, \cdots, x_n)$ is a set of elements, none of which is zero. We say that y is a *least common multiple* of this set of elements if

(LCM1) $x_i \mid y \; \forall i$,
(LCM2) $x_i \mid z \; \forall i$ implies that $y \mid z$.

Using prime factorizations, one can give a simple expression for a l.c.m. that parallels Fact A.3.8.

Fact A.3.10   Suppose $(x_1, \cdots, x_n)$ is a collection of nonzero, nonunit elements. Express each element x_i in terms of its prime factors as in (A.8). Then a l.c.m. of this collection of elements is given by

$$y = \prod_{i=1}^{n} p_j^{\gamma_j} , \tag{A.11}$$

where

$$\gamma_j = \max_i \alpha_{ij} . \tag{A.12}$$

The proof is left as an exercise (see Problem A.3.8).

PROBLEMS

A.3.1. Prove that two elements x, y in a p.i.d. $\mathbf{R}$ are coprime if and only if there exist $p, q \in \mathbf{R}$ such that $px + qy$ is a unit.

A.3.2. Let $\mathbf{R}$ be a domain, and let $\langle x, y \rangle$ denote a GCD of x, y which is unique to within a unit factor. Show that if u, v are units, then $\langle x, y \rangle = \langle ux, vy \rangle$ (in other words, every GCD of x, y is also a GCD of ux, vy and vice versa).

A.3.3. Let $\mathbf{R}$ be a p.i.d., and let $x, y, z \in \mathbf{R}$. Show that $\langle x, y, z \rangle = \langle \langle x, y \rangle, z \rangle$. More generally, show that if $x_1, \cdots, x_n \in \mathbf{R}$, then $\langle x_1, \cdots, x_n \rangle = \langle \langle x_1, \cdots, x_m \rangle, \langle x_{m+1}, \cdots, x_n \rangle \rangle$ for $1 \le m < n$.

A.3.4. Let $\mathbf{R}$ be a p.i.d., and let $x_1, \cdots, x_n \in \mathbf{R}$. Let d be a GCD of this set of elements and select $p_1, \cdots, p_n \in \mathbf{R}$ such that $\sum p_i x_i = d$. Show that 1 is a GCD of the set of elements $p_1, \cdots, p_n$.

A.3.5. Suppose x, y are coprime and z divides y, in a p.i.d. $\mathbf{R}$. Show that x and z are coprime.

A.3.6. Suppose u is a unit and $x_1, \cdots, x_n$ are nonzero elements. Show that $\langle u, x_1, \cdots, x_n \rangle = 1$.

A.3.7. Prove Fact A.3.8.

A.3.8. Prove Fact A.3.10.

A.3.9. (i) Find the prime factorizations of 8, 24, 42 in the ring of integers.

(ii) Find a GCD of the above three numbers using Fact A.3.8.

(iii) Find a l.c.m. of the above three numbers using Fact A.3.10.

A.3.10. Show that the following three statements are equivalent:

(i) x divides y.

(ii) x is a GCD of x, y.

(iii) y is a l.c.m. of x, y.

A.3.11. Suppose $\mathbf{R}$ is a p.i.d., that $x, y, z \in \mathbf{R}$, x and y are coprime, and that x divides yz. Show that x divides z. (Hint: Use Corollary A.3.9.)

A.4   EUCLIDEAN DOMAINS

In this section, we study a special type of ring that finds a lot of application in this book. Throughout this section "domain" means a commutative domain with identity.

Definition A.4.1   A domain $\mathbf{R}$ is a *Euclidean domain* if there is a degree function $\delta : \mathbf{R} \setminus 0 \to \mathbf{Z}_+$ a satisfying the following axioms:[2]

(ED1) For every $x, y \in \mathbf{R}$ with $y \neq 0$, there exists a $q \in \mathbf{R}$ such that either $r := x - qy$ is zero, or else $\delta(r) < \delta(y)$.

(ED2) If $x \mid y$ then $\delta(x) \leq \delta(y)$.

One can think of q as a quotient, and r as a remainder, after "dividing" x by y. The axiom (ED1) states that we can always get a remainder that is either zero or else has a smaller degree than the divisor y. We speak of *a* quotient and remainder because q and r are not necessarily unique unless additional conditions are imposed on the degree function $\delta(\cdot)$.

The axiom (ED2) implies that $\delta(1) \leq \delta(x) \,\forall x \neq 0$, since 1 divides every nonzero element. Hence, it can be assumed without loss of generality that $\delta(1) = 0$. The same axiom also implies that if x and y are associates, then they have the same degree (because if x and y are associates, then $x \mid y$ and $y \mid x$). In particular, $\delta(u) = 0$ whenever u is a unit.

Fact A.4.2   Let $\mathbf{R}$ be a Euclidean domain with degree function $\delta(\cdot)$. and suppose

$$\delta(x + y) \leq \max\{\delta(x), \delta(y)\}, \tag{A.1}$$
$$\delta(xy) = \delta(x) + \delta(y). \tag{A.2}$$

Then, for every $x, y \in \mathbf{R}$ with $y \neq 0$, there exists a *unique* $q \in \mathbf{R}$ such that $\delta(x - yq) < \delta(y)$, where the degree of zero is taken as $-\infty$.

Proof. By (ED1), there exists at least one such q. Now suppose $\delta(x - ay) < \delta(y)$, $\delta(x - by) < \delta(y)$, and define $r = x - ay$, $s = x - by$. Then $x = ay + r = by + s$. Rearranging gives $(a - b)y = s - r$. If $a \neq b$, then $\delta((a - b)y) = \delta(a - b) + \delta(y) \geq \delta(y)$, by (A.2). On the other hand, $\delta(s - r) \leq \max\{\delta(r), \delta(s)\} < \delta(y)$. This contradiction shows that $a = b$ and also $r = s$.   □

Definition A.4.3   A Euclidean domain $\mathbf{R}$ with degree function $\delta(\cdot)$ is called a *proper Euclidean domain* if $\mathbf{R}$ is not a field and $\delta(\cdot)$ satisfies (A.2).[3]

[2]Note that $\mathbf{Z}_+$ denotes the set of nonnegative integers.
[3]This is slightly different from the definition in [65, p. 30].

Note that, in a proper Euclidean domain, the division process might still produce nonunique quotients and remainders, because (A.1) is not assumed to hold. This is the case, for example, in the ring of proper stable rational functions, which are studied in Chapter 2.

Fact A.4.4   Every Euclidean domain is a principal ideal domain.

Proof. Let $\mathbf{R}$ be a Euclidean domain, and let $\mathbf{I}$ be an ideal in $\mathbf{R}$. If $\mathbf{I} = \{0\}$, then $\mathbf{I}$ is principal with 0 as the generator. So suppose $\mathbf{I}$ contains some nonzero elements, and let x be an element of $\mathbf{I}$ such that $\delta(x)$ is minimum over all nonzero elements of $\mathbf{I}$. We claim that $\mathbf{I}$ is the ideal generated by x and is hence principal. To prove this, let $y \in \mathbf{I}$ be chosen arbitrarily; it is shown that x divides y. By axiom (ED1), there exists a $q \in \mathbf{R}$ such that either $r := y - qx$ is zero or else $\delta(r) < \delta(x)$. If $r \neq 0$, then $\delta(r) < \delta(x)$ contradict's the manner in which x was chosen. Hence, $r = 0$, i.e., x divides y.
□

We now present a very important example of a Euclidean domain, namely the ring of polynomials in one indeterminate with coefficients in a field. To lead up to this example, an abstract definition of a polynomial is given.

Let $\mathbf{R}$ be a ring. Then a *polynomial* over $\mathbf{R}$ is an infinite sequence $\{a_0, a_1, \cdots\}$ such that only finitely many terms are nonzero. The sum and product of two polynomials $a = \{a_i\}$ and $b = \{b_i\}$ are defined by

$$(a + b)_i = a_i + b_i \, , \tag{A.3}$$

$$(ab)_i = \sum_{j=0}^{i} a_{i-j} \, b_j = \sum_{j=0}^{i} a_j \, b_{i-j} \, . \tag{A.4}$$

For notational convenience, a polynomial $a = \{a_i\}$ can be represented by $a_0 + a_1 s + a_2 s^2 + \cdots$ where s is called the "indeterminate." The highest value of the index i such that $a_i \neq 0$ is called the *degree* of a polynomial $a = \{a_0, a_1, \cdots\}$.[4] Thus, if a is a polynomial of degree m, we can write

$$a(s) = a_0 + a_1 s + \cdots + a_m s^m = \sum_{i=0}^{m} a_i s^i \, . \tag{A.5}$$

The set of polynomials over $\mathbf{R}$ is denoted by $\mathbf{R}[s]$, and is a commutative ring with identity. Moreover, if $\mathbf{R}$ is a domain, so is $\mathbf{R}[s]$.

Fact A.4.5   Suppose $\mathbf{F}$ is a field. Then $\mathbf{F}[s]$ is a Euclidean domain if the degree of a polynomial in $\mathbf{F}[s]$ is defined as above.

---

[4]The degree of the zero polynomial is taken as $-\infty$.

Proof. To prove axiom (ED1), suppose

$$f(s) = \sum_{i=0}^{n} f_i \, s^i, \quad g(s) = \sum_{i=0}^{m} g_i \, s^i, g_m \neq 0 \,. \tag{A.6}$$

It is necessary to show the existence of a $q \in \mathbf{F}[s]$ such that $\delta(f - gq) < \delta(g)$, where $\delta(0) = -\infty$. If $n < m$, take $q = 0$. If $n \geq m$, define $q_1(s) = (f_n/g_m)s^{n-m}$; then $\delta(f - gq_1) \leq n - 1$. By repeating this process if necessary on the polynomial $f - gq_1$ we can ultimately find a $q \in \mathbf{F}[s]$ such that $\delta(f - gq) < \delta(g)$. Thus, (ED1) is satisfied. The proof of (ED2) is straight-forward. □

Since the degree function $\delta(\cdot)$ satisfies both (A.1) and (A.2), the Euclidean division process yields a *unique* remainder and quotient r, q corresponding to each pair f, g with $g \neq 0$. Moreover, it is clear that $\mathbf{F}[s]$ is not a field (the polynomial s has no inverse). Hence, $\mathbf{F}[s]$ is a proper Euclidean domain.

The field of fractions associated with $\mathbf{F}[s]$ is denoted by $\mathbf{F}(s)$, and is called the set of *rational functions* over $\mathbf{F}$. Note that every element of $\mathbf{F}(s)$ is a ratio of two polynomials (hence the name rational function).

PROBLEMS

A.4.1. Show that the set of integers $\mathbf{Z}$ is a Euclidean domain if we define the degree of an integer to be its absolute value. Does the division process result in unique remainders?

A.4.2. Consider the ring $\mathbb{R}[s]$, consisting of polynomials with real coefficients. What are the primes of this ring?

A.4.3. Suppose $\mathbf{R}$ is a proper Euclidean domain and $x \in \mathbf{R}$. Show that if $\delta(x) = 1$, then x is a prime. Is the converse true? (Hint: See Problem A.4.2.)

A.4.4. Let $\mathbf{R}$ be a Euclidean domain. Show that $x \in \mathbf{R}$ is a unit if and only if $\delta(x) = 0$. (Hint: Use axiom (ED1).)

APPENDIX B

Preliminaries on Matrix Rings

The objective of this appendix is to gather some well-known facts on matrices whose elements belong to a ring or a field, and to state them in as much generality as possible.

B.1   MATRICES AND DETERMINANTS

Let $\mathbf{R}$ be a ring, and let $\mathbf{R}^{n \times n}$ denote the set of *square* matrices of order $n \times n$ whose elements belong to $\mathbf{R}$. If the sum and product of two matrices in $\mathbf{R}^{n \times n}$ are defined in the familiar way, namely

$$(A + B)_{ij} = a_{ij} + b_{ij} \,, \tag{B.1}$$

$$(AB)_{ij} = \sum_{k=1}^{n} a_{ik} b_{kj} \,, \tag{B.2}$$

then $\mathbf{R}^{n \times n}$ becomes a ring, usually referred to as a *ring of matrices* over $\mathbf{R}$. If $\mathbf{R}$ contains an identity and $n \geq 2$, then $\mathbf{R}^{n \times n}$ is not commutative. For instance, if $n = 2$, we have

$$\begin{bmatrix} 1 & 0 \\ 0 & 0 \end{bmatrix} \begin{bmatrix} 0 & 0 \\ 1 & 0 \end{bmatrix} \neq \begin{bmatrix} 0 & 0 \\ 1 & 0 \end{bmatrix} \begin{bmatrix} 1 & 0 \\ 0 & 0 \end{bmatrix}. \tag{B.3}$$

Similar examples can be constructed if $n > 2$. Also, if $n \geq 2$, then $\mathbf{R}^{n \times n}$ is not a domain because

$$\text{Diag} \{1, 0, \cdots, 0\} \, \text{Diag} \{0, 1, 0, \cdots, 0\} = 0_{n \times n} \,. \tag{B.4}$$

The *determinant* of a matrix $A \in \mathbf{R}^{n \times n}$ is denoted by $|A|$ and is defined in the familiar way, namely

$$|A| = \sum_{\phi \in \Pi_n} \text{sign } \phi \prod_{i=j}^{n} a_{i\phi(i)} \,, \tag{B.5}$$

where Π_n denotes the set of permutations of the set $\mathbf{N} = \{1, \cdots, n\}$ into itself, and sign $\phi = \pm 1$ depending on whether ϕ is an even or odd permutation.[1] Most of the usual properties of determinants hold in the present abstract setting. The required results are stated without proof, and the reader is referred to [65] for further details.

Define a function $\Delta : \mathbf{R}^n \times \cdots \times \mathbf{R}^n \to \mathbf{R}$ as follows: For every $v_1, \cdots, v_n \in \mathbf{R}^n$, define $\Delta(v_1, \cdots, v_n)$ to be the determinant of the matrix $V \in \mathbf{R}^{n \times n}$ whose columns are $v_1, \cdots, v_n$ in that

---

[1]Throughout this section, the symbol "1" is used to denote both the integer as well as the identity element of the ring $\mathbf{R}$. It is usually clear from the context which is meant.

order. Thus, Δ is just the determinant function viewed as a function of the columns of a matrix. Similarly, define $\bar{\Delta}(v_1, \cdots, v_n)$ to be the determinant of the matrix $V' \in \mathbf{R}^{n \times n}$ whose *rows* are $v_1, \cdots, v_n$, in that order.

Fact B.1.1   We have

$$\Delta(v_1, \cdots, v_n) = \bar{\Delta}(v_1, \cdots, v_n) \, \forall \, v_1, \cdots, v_n \in \mathbf{R}^n \, . \tag{B.6}$$

The function Δ (and hence $\bar{\Delta}$) is alternating and multilinear. That is, if two arguments of Δ are interchanged, then Δ changes sign, and if two arguments of Δ are equal, then Δ equals zero; finally,

$$\Delta(\alpha v_1 + \beta w_1, v_2, \cdots, v_n) = \alpha \Delta(v_1, v_2, \cdots, v_n) + \beta \Delta(w_1, v_2, \cdots, v_n) \tag{B.7}$$

for all possible choices of the arguments.

Fact B.1.2   Let $A \in \mathbf{R}^{n \times n}, n \geq 2$. Then, for any $i, j \in \{1, \cdots, n\}$,

$$|A| = \sum_{j=1}^{n} (-1)^{i+j} a_{ij} \, m_{ij}(A) \, , \tag{B.8}$$

$$|A| = \sum_{i=1}^{n} (-1)^{i+j} a_{ij} \, m_{ij}(A) \, , \tag{B.9}$$

where $m_{ij}(A)$ is the ij-*th minor* of A, defined as the determinant of the $(n-1) \times (n-1)$ matrix obtained from A by deleting its i-th row and j-th column.[2]

Fact B.1.3   Let $A \in \mathbf{R}^{n \times n}, n \geq 2$. Then

$$\sum_{j=1}^{n} a_{ij} (-1)^{k+j} \, m_{kj}(A) = \begin{cases} |A| & \text{if } i = k \\ 0 & \text{if } i \neq k \end{cases} \tag{B.10}$$

$$\sum_{i=1}^{n} a_{ik} (-1)^{i+j} \, m_{ij}(A) = \begin{cases} |A| & \text{if } j = k \\ 0 & \text{if } j \neq k \end{cases} . \tag{B.11}$$

Now a bit of notation is introduced to make subsequent theorem statements more compact. Suppose m and n are positive integers, with $m \leq n$. Then $S(m, n)$ denotes the collection of all strictly increasing m-tuples $\{i_1, \cdots, i_m\}$, where $1 \leq i_1 < i_2 < \cdots < i_m \leq n$. For example,

$$S(3, 5) = \{(1, 2, 3), (1, 2, 4), (1, 2, 5), (1, 3, 4), (1, 3, 5) \, ,$$
$$(1, 4, 5), (2, 3, 4), (2, 3, 5), (2, 4, 5), (3, 4, 5)\} \, . \tag{B.12}$$

---

[2]To be consistent with subsequent notation, one should write $a_{N \setminus i, N \setminus j}$ instead of $m_{ij}(A)$. But the latter notation is more convenient.

If $m = n$, then $S(m, n)$ is the singleton set $\{(1, 2. \cdots , n)\}$, while if $m = 0$ then $S(m, n)$ is just the empty set.

Suppose $A \in \mathbf{R}^{m \times n}$, and let $I \in S(l, m)$, $J \in S(l, n)$. Then a_{IJ} denotes the $l \times l$ minor of A consisting of the rows from I and the columns from J. In particular, if I and J are singleton sets of the form $I = \{i\}$, $J = \{j\}$, then a_{IJ} is just the element a_{ij}. Note that $a_{IJ} \in \mathbf{R}$. We use A_{IJ} to denote the $l \times l$ *matrix* consisting of the elements from the rows in I and the columns in J. Thus, $a_{IJ} = |A_{IJ}|$. In some situations it is of interest to examine a submatrix of A consisting of the rows in I and *all* columns of A; such a submatrix is denoted by $A_{I.}$. The notation $A_{.J}$ is similarly defined.

Fact B.1.4   (Laplace's Expansion of a Determinant) Suppose $A \in \mathbf{R}^{n \times n}$, and suppose $I \in S(m, n)$. Then,

$$|A| = \sum_{J \in S(m,n)} (-1)^{v(I,J)} a_{IJ}\, a_{\mathbf{N}\setminus I, \mathbf{N}\setminus J} \,, \tag{B.13}$$

where

$$v(I, J) = \sum_{i \in I} i + \sum_{j \in J} j \,. \tag{B.14}$$

Equation (B.13) generalizes (B.8). The corresponding generalization of (B.9) is similar and is left to the reader.

Fact B.1.5   (Binet-Cauchy Formula) Suppose $A \in \mathbf{R}^{n \times m}$, $B \in \mathbf{R}^{m \times l}$, and let $C = AB \in \mathbf{R}^{n \times l}$. Let $I \in S(p, n)$, $J \in S(p, l)$. Then,

$$c_{IJ} = \sum_{K \in S(p,m)} a_{IK}\, b_{KJ} \,. \tag{B.15}$$

In particular, if $n = m = l$, then $|C| = |A| \cdot |B|$.

Note that (B.15) is a natural generalization of (B.2).

Using the multilinearity of the determinant function, one can obtain an expression for the determinant of the sum of two matrices. For example, if $A, B \in \mathbf{R}^{2 \times 2}$ and a_1, a_2, b_1, b_2 are the columns of the two matrices, then

$$\begin{aligned}|A + B| &= \Delta(a_1 + b_1, a_2 + b_2) \\ &= \Delta(a_1, a_2) + \Delta(a_1, b_2) + \Delta(b_1, a_2) + \Delta(b_1, b_2) \,. \end{aligned} \tag{B.16}$$

If $A, B \in \mathbf{R}^{n \times n}$ then the formula for $|A + B|$ will involve the sum of 2^n terms. In case one of the matrices is diagonal, each term in this expansion can be neatly expressed as a product of principal minors of A and B.

Fact B.1.6   Suppose $A, B \in \mathbf{R}^{n \times n}$ and that A is diagonal (i.e., $a_{ij} = 0$ for $i \neq j$). Then,

$$|A + B| = \sum_{l=0}^{n} \sum_{I \in S(l,n)} a_{II}\, b_{\mathbf{N}\setminus I, \mathbf{N}\setminus I} \,, \tag{B.17}$$

where a_{II} is interpreted as 1 when I is empty.

If $A \in \mathbf{R}^{n \times n}$, its *adjoint matrix*, denoted by A^{adj}, is defined by

$$(A^{adj})_{ij} = (-1)^{i+j} m_{ji}(A) \ . \tag{B.18}$$

In view of (B.10) and (B.11), it is seen that, for any $A \in \mathbf{R}^{n \times n}$,

$$A \cdot A^{adj} = A^{adj} \cdot A = |A| I_n \ , \tag{B.19}$$

where I_n denotes the $n \times n$ identity matrix.

A matrix $A \in \mathbf{R}^{n \times n}$ ts *unimodular* if it has an inverse in $\mathbf{R}^{n \times n}$, i.e., it is a unit in the ring $\mathbf{R}^{n \times n}$.

Fact B.1.7 $A \in \mathbf{R}^{n \times n}$ is unimodular if and only if $|A|$ is a unit in $\mathbf{R}$.

Proof. "if" Suppose $|A|$ is a unit in $\mathbf{R}$ and let $b = |A|^{-1} \in \mathbf{R}$. Then $bA^{adj} \in \mathbf{R}^{n \times n}$ and $A \cdot bA^{adj} = bA^{adj} \cdot A = I_n$. Hence, A is unimodular.

"only if" Suppose A is unimodular, and let $B \in \mathbf{R}^{n \times n}$ be the inverse of A. Then $1 = |I_n| = |A| \cdot |B|$, which shows that $|A|$ is a unit in $\mathbf{R}$. $\qquad \square$

Now consider the set $\mathbf{F}^{n \times n}$ of matrices with elements in a *field* $\mathbf{F}$. Since a field is also a ring, all of the preceding discussion applies. In addition, since every nonzero element of $\mathbf{F}$ is a unit, we see that every $A \in \mathbf{F}^{n \times n}$ such that $|A| \neq 0$ has an inverse in $\mathbf{F}^{n \times n}$. It is customary to call a matrix A *nonsingular* if $|A| \neq 0$. From Fact B.1.3, we see that if A is nonsingular, then A^{-1} is given by

$$(A^{-1})_{ij} = (-1)^{i+j} m_{ji}(A)/|A| \ . \tag{B.20}$$

The relation (B.20) is a special case of the following result, which gives the relationship between the minors of A and A^{-1}. The proof can be found in [41, pp. 21–22].

Fact B.1.8 Suppose $A \in \mathbf{F}^{n \times n}$ is nonsingular and let $B = A^{-1}$. Let $\mathbf{N} = \{1, \cdots, n\}$ and suppose $I, J \in S(l, n)$ for some l. Then,

$$b_{IJ} = (-1)^{v(\mathbf{N} \backslash I, \mathbf{N} \backslash J)} a_{\mathbf{N} \backslash I, \mathbf{N} \backslash J} |A|^{-1} \ , \tag{B.21}$$

where the function v is defined in (B.14).

If I and J are singleton sets then (B.21) reduces to (B.20).

Suppose $F \in \mathbf{F}^{r \times s}$, and suppose F can be partitioned as

$$F = \begin{bmatrix} A & B \\ C & D \end{bmatrix}, \tag{B.22}$$

where A is a nonsingular matrix of order $n \times n$. Then the matrix

$$G := D - CA^{-1}B \in \mathbf{F}^{r-n \times s-n} \tag{B.23}$$

is called the *Schur complement* of F with respect to A, and is sometimes denoted by F/A. The next result relates the minors of G to those of F.

Fact B.1.9   Suppose $J \in S(t, m)$, $K \in S(t, l)$. Then

$$g_{JK} = |A|^{-1} f_{\mathbf{N} \cup (\{n\}+J), \mathbf{N} \cup (\{n\}+K)}, \tag{B.24}$$

where $\{n\} + J$ denotes the set sum (thus if $J = \{j_1, \cdots, j_t\}$, then $\{n\} + J = \{n + j_1, \cdots, n + j_t\}$).

Proof. Observe that

$$\begin{bmatrix} I & 0 \\ -CA^{-1} & I \end{bmatrix} F = \begin{bmatrix} A & B \\ 0 & G \end{bmatrix} =: E \ . \tag{B.25}$$

Now suppose $P \in S(n + t, n + m)$, $Q \in S(n + t, n + l)$ and that $\mathbf{N}$ is a subset of both P and Q. Consider the minor e_{PQ}. The rows in P of E are obtained by adding multiples of the first n rows of F to the rows in P of F. Thus, $e_{PQ} = f_{PQ}$, since a minor is unchanged by adding multiples of some rows to others. Observe that the matrix E_{PQ} is block-lower-triangular, so that $e_{PQ} = |A| \cdot e_{P-\mathbf{N}, Q-\mathbf{N}}$. The identity (B.24) now follows by choosing $P = \mathbf{N} \cup (\{n\} + J)$, $Q = \mathbf{N} \cup (\{n\} + K)$. $\qquad \square$

Fact B.1.10   Suppose $A \in \mathbf{F}^{n \times n}$, $B \in \mathbf{F}^{m \times n}$, $|A| \neq 0$, and let $G = BA^{-1}$. Suppose $J \in S(l, m)$, $K \in S(l, n)$. Then $|A| \cdot g_{JK}$ equals the determinant of the $n \times n$ matrix obtained from A by replacing the rows in K of A by the rows in J of B.

Remarks B.1.11   A simple example helps to illustrate the statement of the result. Suppose $A \in \mathbf{F}^{5 \times 5}$, $B \in \mathbf{F}^{3 \times 5}$, and let a^i, b^i denote the i-th rows of A and B, respectively. Each of these is a 1×5 row vector. Suppose $J = (2, 3)$, $K = (3, 5)$. Then the above result states that

$$|A| \cdot g_{(2,3),(3,5)} = \begin{vmatrix} a^1 \\ a^2 \\ b^2 \\ a^4 \\ b^3 \end{vmatrix} . \tag{B.26}$$

Note that if B is a column vector then the above fact reduces to Cramer's rule for solving linear equations.

Proof. Define

$$F = \begin{bmatrix} A & -I \\ B & 0 \end{bmatrix}.$$ (B.27)

Then G is the Schur complement of F with respect to A. By Fact B.1.9,

$$|A| \cdot g_{JK} = f_{\mathbf{N} \cup (\{n\} + J)}, \mathbf{N} \cup (\{n\} + K),$$

$$= \begin{vmatrix} A & -I_{.K} \\ B_{J.} & 0 \end{vmatrix}.$$ (B.28)

Note that each of the last l columns of the minor consists of all zeros except for a single 1. If we expand this minor about the last l columns using Laplace's expansion (Fact B.1.4), then the expansion consists of a single term, since the submatrix consisting of the last l columns has only nonzero $l \times l$ minor. Thus, if $K = \{k_1, \cdots, k_l\}$, then

$$|A| \cdot g_{JK} = (-1)^v (-1)^l \begin{vmatrix} A_{\mathbf{N} \setminus K} \\ B_{J.} \end{vmatrix},$$ (B.29)

where

$$v = \sum_{i=1}^{l} (n + i) + \sum_{i=1}^{l} k_i .$$ (B.30)

Now the minor on the right side of (B.29) is not quite in the form stated in Fact B.1.10, since the rows of B occur *below* those of A, rather than *substituting* for them (as in Example B.26, for instance). It is a matter of detail to verify that the parity factor $(-1)^{v+l}$ accounts for this difference.
□

Fact B.1.12   Suppose $A \in \mathbf{F}^{n \times n}$, $B \in \mathbf{F}^{n \times m}$, $|A| \neq 0$ and let $G = A^{-1} B$. Suppose $J \in S(l, n)$, $K \in S(l, m)$. Then $|A| \cdot g_{JK}$ equals the determinant of the $n \times n$ matrix obtained from A by replacing the columns in J of A by the columns in K of B.

The proof is similar to that of Fact B.1.10 and is left to the reader.

B.2   CANONICAL FORMS

Let $\mathbf{R}$ be a principal ideal domain, and let $\mathbf{F}$ be the field of fractions associated with $\mathbf{R}$. In this section, we study the sets $\mathbf{R}^{n \times m}$ and $\mathbf{F}^{n \times m}$, consisting of $n \times m$ matrices whose elements belong to $\mathbf{R}$ and $\mathbf{F}$, respectively. We prove the existence of two canonical forms, namely the *Smith form* on $\mathbf{R}^{n \times m}$ and the *Smith-McMillan form* on $\mathbf{F}^{n \times m}$.

A matrix $A \in \mathbf{R}^{n \times m}$ is a *left associate* of $B \in \mathbf{R}^{n \times m}$ (denoted by $A =_L B$) if there is a unimodular matrix $U \in \mathbf{R}^{n \times n}$ such that $A = UB$. A is a *right associate* of B (denoted by $A =_R B$) if there is a unimodular matrix $V \in \mathbf{R}^{m \times m}$ such that $A = BV$. A is *equivalent* to B (denoted by $A \sim B$) if there are unimodular matrices $U \in \mathbf{R}^{n \times n}$, $V \in \mathbf{R}^{m \times m}$. such that $A = UBV$. It is left to the reader to verify that $=_L$, $=_R$, $\sim$ are all equivalence relations.

We now commence our study of canonical forms.

Lemma B.2.1 *Suppose $a_1, \cdots, a_n \in \mathbf{R}$, and let d_n be a GCD of this set of elements. Then there exists a matrix $P_n \in \mathbf{R}^{n \times n}$ whose first row is $[a_1 \cdots a_n]$ and whose determinant is d_n.*

Proof. The proof is by induction on n. If $n = 2$, by Theorem A.3.3 there exist $p_1, p_2 \in \mathbf{R}$ such that $p_1 a_1 + p_2 a_2 = d_2$. Now let

$$P_2 = \begin{bmatrix} a_1 & a_2 \\ -p_2 & p_1 \end{bmatrix} \tag{B.1}$$

For larger values of n, let d_{n-1} be a GCD of $a_1, \cdots, a_{n-1}$. By the inductive hypothesis, we can construct a matrix $P_{n-1} \in \mathbf{R}^{n-1 \times n-1}$ whose first row is $[a_1 \cdots a_{n-1}]$ and whose determinant is d_{n-1}. By Fact B.1.3,

$$\sum_{j=1}^{n-1} a_j (-1)^{1+j} m_{1j}(P_{n-1}) = d_{n-1} . \tag{B.2}$$

so that

$$\sum_{j=1}^{n-1} \frac{a_j}{d_{n-1}} (-1)^{1+j} m_{1j}(P_{n-1}) = 1 . \tag{B.3}$$

For convenience, let $z_j = -a_j/d_{n-1}$. Next, by Problem A.3.3, d_n is a GCD of d_{n-1} and a_n. By Theorem A.3.3, there exist $x, y \in \mathbf{R}$ such that $xd_{n-1} + ya_n = d_n$. Now define

$$P_n = \begin{bmatrix} & & & a_n \\ & P_{n-1} & & 0 \\ & & & \vdots \\ & & & 0 \\ yz_1 & \cdots & yz_{n-1} & x \end{bmatrix} . \tag{B.4}$$

Expanding $|P_n|$ about the last column and using (B.3) gives $|P_n| = xd_{n-1} + ya_n = d_n$. □

Theorem B.2.2 (Hermite Form) Every $A \in \mathbf{R}^{n \times n}$ is a left associate of a matrix B that is lower-triangular (i.e., $b_{ij} = 0$ for $j > i$).

Proof. Let d_n be a GCD of $\{a_{1n}, \cdots, a_{nn}\}$, i.e., the elements of the last column of A. By Theorem A.3.5, there exist elements $p_1, \cdots, p_n \in \mathbf{R}$ such that

$$\sum_{i=1}^{n} p_i\, a_{in} = d_n\ . \tag{B.5}$$

By Problem A.3.4, 1 is a GCD of the set $p_1, \cdots, p_n$. Hence, by a slight variation of Lemma B.2.1, there exists a unimodular matrix $U \in \mathbf{R}^{n \times n}$ such that its *last* row is $[p_1 \cdots p_n]$. Now $(UA)_{nn} = d_n$. Also, for $i = 1, \cdots, n-1$, $(UA)_{in}$ belongs to the ideal generated by $a_{1n}, \cdots, a_{nn}$ and is thus a multiple of d_n. Let $z_i = (UA)_{in}/d_n$ for $i = 1, \cdots, n-1$, and define

$$U_n = \begin{bmatrix} 1 & 0 & \cdots & 0 & -z_1 \\ 0 & 1 & \cdots & 0 & -z_2 \\ \vdots & \vdots & \vdots & \vdots & \vdots \\ 0 & 0 & \cdots & 1 & -z_{n-1} \\ 0 & 0 & \cdots & 0 & 1 \end{bmatrix}. \tag{B.6}$$

Then U_n is unimodular since its determinant is 1. Moreover, $U_n U A$ is of the form

$$U_n U A = \begin{bmatrix} & & 0 \\ A_{n-1} & & \vdots \\ & & 0 \\ a^{n-1} & & d_n \end{bmatrix}. \tag{B.7}$$

In other words, the last column of A has been reduced to zero above the diagonal by means of left multiplication by an appropriate unimodular matrix. One can now repeat the procedure with the $n-1 \times n-1$ matrix A_{n-1}, and eventually arrive at a lower triangular matrix. For clarity, the next step of the algorithm is briefly outlined: Let d_{n-1} denote a GCD of the elements of the last column of A_{n-1}. As above, there exists a unimodular matrix $\bar{U}_{n-1} \in \mathbf{R}^{n-1 \times n-1}$ such that

$$\bar{U}_{n-1}\, A_{n-1} = \begin{bmatrix} & & 0 \\ A_{n-2} & & \vdots \\ & & 0 \\ a^{n-2} & & d_{n-1} \end{bmatrix}. \tag{B.8}$$

Now define the $n \times n$ unimodular matrix

$$U_{n-1} = \text{Block Diag}\{\bar{U}_{n-1}, 1\}\ . \tag{B.9}$$

Then,

$$U_{n-1} U_n U A = \begin{bmatrix} & & 0 & 0 \\ A_{n-2} & & \vdots & \vdots \\ & & 0 & 0 \\ a^{n-2} & d_{n-1} & 0 \\ a^{n-1} & \cdot & d_n \end{bmatrix}. \tag{B.10}$$

The rest of the proof is now obvious.   □

Corollary B.2.3   *Every matrix $A \in \mathbf{R}^{n \times n}$, is a left associate of an upper triangular matrix $C \in \mathbf{R}^{n \times n}$.*

Proof. In the proof of Theorem B.2.2, start with the first column of A instead of the last.   □

Corollary B.2.4   *Every matrix $A \in \mathbf{R}^{n \times n}$ is a right associate of a lower (resp. upper) triangular matrix.*

Proof. In the proof of Theorem B.2.2, start with the last (resp. first) row of A.   □

For rectangular matrices, the following result holds:

Corollary B.2.5   *Suppose $A \in \mathbf{R}^{n \times m}$. Then A is a left associate of a matrix of the form*

$$\begin{bmatrix} D \\ 0 \end{bmatrix} \text{ if } n > m, \ [D \ E] \text{ if } n < m . \tag{B.11}$$

where D can be chosen to be either lower or upper triangular.

It is left to the reader to state and prove the result analogous to Corollary B.2.5 concerning right associates of rectangular matrices.

Next, we start developing the Smith form. Recall that a matrix $A \in \mathbf{R}^{n \times m}$ is said to have *rank* l if (i) there is an $l \times l$ submatrix of A with nonzero determinant, and (ii) every $(l + 1) \times (l + 1)$ minor of A is zero. An *elementary row operation* on the matrix A consists of one of the following: (i) interchanging two rows of A, or (ii) adding a multiple of one row to another. An *elementary column operation* is similarly defined. It is easy to see that an elementary row (column) operation on a matrix A can be accomplished by multiplying A on the left (right) by a unimodular matrix. Thus, a matrix obtained from A by elementary row and column operations is equivalent to A.

Now suppose $A \in \mathbf{R}^{n \times m}$ is a (left or right) multiple of $B \in \mathbf{R}^{n \times m}$, and let b_l denote a GCD of all $l \times l$ minors of B. Then it follows from the Binet-Cauchy formula (Fact B.1.5) that b_l divides all $l \times l$ minors of A. Thus, if a_l denotes a GCD of all $l \times l$ minors of A, we see that b_l divides a_l.

From this, it follows that if A and B are equivalent matrices, then a_l and b_l are associates, and A and B have the same rank.

Theorem B.2.6   (Smith Form) Suppose $A \in \mathbf{R}^{n \times m}$ has rank l. Then, A is equivalent to a matrix $H \in \mathbf{R}^{n \times m}$ of the form

$$
H = \begin{bmatrix}
h_1 & 0 & \cdots & 0 & 0 \\
0 & h_2 & \cdots & 0 & 0 \\
\vdots & \vdots & \vdots & \vdots & \vdots \\
0 & 0 & 0 & h_l & 0 \\
 & & 0 & & 0
\end{bmatrix}, \tag{B.12}
$$

where h_i divides h_{i+1} for $i = 1, \cdots, l - 1$. Moreover, $h_1 \cdots h_i$ is a GCD of all $i \times i$ minors of A, and the h_i's are unique to within multiplication by a unit.

Remarks B.2.7 $h_1, \cdots, h_n$ are called the *invariant factors* of A.

Proof. Since A has rank l, it contains an $l \times l$ submatrix with nonzero determinant. By elementary row and column operations, this submatrix can be brought to the upper left-hand corner of A.

As in the proof of Theorem B.2.2, there is a unimodular matrix $U \in \mathbf{R}^{n \times n}$ such that $(UA)_{11}$ is a GCD of all elements in the first column of A. By elementary row operations, the first column of UA can be made to contain all zeros except in the $(1, 1)$-position. Call the resulting matrix $\bar{A}$. If $\bar{a}_{11}$ divides all elements of the first row of $\bar{A}$, then all elements of the first row (except in the $(1, 1)$ position) of $\bar{A}V$ can be made to equal zero by a suitable choice of a unimodular matrix V of the form

$$
V = \begin{bmatrix}
1 & -v_{12} & -v_{13} & \cdots & -v_{1m} \\
\cdot & 1 & 0 & \cdot & 0 \\
\cdot & \cdot & 1 & \cdot & 0 \\
\cdot & \cdot & \cdot & \vdots & \vdots \\
\cdot & \cdot & \cdot & \cdot & 1
\end{bmatrix}. \tag{B.13}
$$

Moreover, the first column or $\bar{A}V$ will continue to have zeros except in the $(1, 1)$ position. On the other hand, if $\bar{a}_{11}$ does not divide all elements in the first row of $\bar{A}$. we can choose a unimodular matrix V so that $(\bar{A}V)_{11}$ is a GCD of the elements of the first row of $\bar{A}V$. If this is done, the first *column* of $\bar{A}V$ may no longer contain zeros. In such a case, we repeat the above row and column operations. This process cannot continue indefinitely, because the original element $\bar{a}_{11}$ has only a finite number of prime factors (see Fact A.3.7) and each successive corner element is a proper (i.e., nonassociative) divisor of its predecessor. Thus, in a finite number of steps, we arrive at a matrix B equivalent to A such that b_{11} divides all elements of the first row as well as first column of B. By elementary row and column operations, all of the elements of the first row and column of B can be made equal to zero. Thus,

$$A \sim B \sim \begin{bmatrix} b_{11} & 0 & \cdots & 0 \\ 0 & & & \\ \vdots & & B_1 & \\ 0 & & & \end{bmatrix}. \tag{B.14}$$

By proceeding to the first row and column of B_1 (i.e., the second row and column of B) and then repeating the procedure, we will eventually have

$$A \sim \begin{bmatrix} d_1 & \cdots & 0 & \\ \vdots & \vdots & \vdots & 0 \\ 0 & \cdots & d_l & \\ & 0 & & M \end{bmatrix}. \tag{B.15}$$

Now, $d_i \neq 0 \ \forall i$, since these are all divisors of the elements of the first l rows and columns of A; and none of these rows nor columns is identically zero. Next, M must be zero; otherwise A is equivalent to a matrix of rank at least $l+1$, which contradicts the fact that the rank of A is l.

Thus, far we have shown that

$$A \sim \begin{bmatrix} d_1 & 0 & \cdots & 0 & \\ 0 & d_2 & \cdots & 0 & 0 \\ \vdots & \vdots & \vdots & \vdots & \\ 0 & 0 & \cdots & d_l & \\ & & 0 & & 0 \end{bmatrix}. \tag{B.16}$$

Let $D = \mathrm{Diag}\,\{d_1, \cdots, d_l\}$. We will show that $D \sim \mathrm{Diag}\,\{h_1, \cdots, h_l\}$, where h_i divides h_{i+1} for $i = 1, \cdots, l - 1$. This is enough to prove the theorem. By adding columns 2 to l to the first column of D, it follows that

$$D \sim \begin{bmatrix} d_1 & 0 & \cdots & 0 \\ d_2 & d_2 & \cdots & 0 \\ \vdots & \vdots & \vdots & \vdots \\ d_l & 0 & \cdots & d_l \end{bmatrix}. \tag{B.17}$$

By multiplying the latter matrix on the left by an appropriate unimodular matrix, we get another matrix E whose $(1, 1)$-element is a GCD of $\{d_1, \cdots, d_l\}$ and whose first column is zero otherwise. Let $h_1 := $ a GCD of $\{d_1. \cdots, d_l\}$. Then,

$$E = \begin{bmatrix} h_1 & & \\ 0 & \bar{E} & \\ \vdots & & \\ 0 & & \end{bmatrix}, \tag{B.18}$$

where every element of $\bar{E}$ is in the ideal generated by $d_2, \cdots, d_l$, and is thus a multiple of h_1. Since the first row of $\bar{E}$ is a multiple of h_1, it follows that

$$E \sim \begin{bmatrix} h_1 & 0 & \cdots & 0 \\ 0 & & & \\ \vdots & & E_1 & \\ 0 & & & \end{bmatrix}, \tag{B.19}$$

where every element of E_1 is a multiple of h_1. Now, by the preceding paragraph, $E_1 \sim$ Diag $\{g_2, \cdots, g_l\}$ where h_1 divides g_i for all i. Now let h_2 be a GCD of $g_2, \cdots, g_l$. Then clearly h_1 divides h_2. Moreover,

$$E_1 \sim \begin{bmatrix} h_2 & 0 \\ 0 & E_2 \end{bmatrix}. \tag{B.20}$$

Repeating this procedure, we finally get $D \sim$ Diag $\{h_1, \cdots, h_l\}$ where h_i divides h_{i+1} for all i. This completes the proof of (B.12).

The divisibility conditions on the h_i imply that the product $h_1 \cdots h_i$ is a GCD of all $i \times i$ minors of H. Since A and H are equivalent, product is also a GCD of all $i \times i$ minors of A.

Now suppose A is also equivalent to another matrix

$$G = \begin{bmatrix} g_1 & 0 & \cdots & 0 & \\ 0 & g_2 & \cdots & 0 & 0 \\ \vdots & \vdots & \vdots & \vdots & \\ 0 & 0 & \cdots & g_l & \\ & & 0 & & 0 \end{bmatrix}, \tag{B.21}$$

where g_i divides g_{i+1} for all i. Then the reasoning of the preceding paragraph shows that the product $h_1 \cdots h_i$ is an associate of the product $g_1 \cdots g_i$ for all i. Now, if $h_1 \sim g_1$ and $h_1 h_2 \sim g_1 g_2$, then $h_2 \sim g_2$. Reasoning inductively in this fashion, we conclude that $h_i \sim g_i$ for all i. □

Corollary B.2.8   *Two matrices $A, B \in \mathbf{R}^{n \times m}$ are equivalent if and only if their invariant factors are associates.*

Let $\mathbf{F}$ be the field of fractions associated with $\mathbf{R}$. We now begin a study of the set $\mathbf{F}^{n\times m}$ of matrices with elements from $\mathbf{F}$ and show the existence of a canonical form known as the Smith-McMillan form.

Theorem B.2.9   (Smith-McMillan Form) Suppose $F \in \mathbf{F}^{n\times m}$ has rank l. Then there exist unimodular matrices $U \in \mathbf{R}^{n\times n}$, $V \in \mathbf{R}^{m\times m}$ such that

$$
UFV = \begin{bmatrix}
a_1/b_1 & 0 & \cdots & 0 & 0 \\
0 & a_2/b_2 & \cdots & 0 & 0 \\
\vdots & \vdots & \vdots & \vdots & \vdots \\
0 & 0 & \cdots & a_l/b_l & \\
& & & 0 & 0
\end{bmatrix},
\tag{B.22}
$$

where a_i, b_i are coprime for all i; a_i divides a_{i+1} and b_{i+1} divides b_i for $i = 1, \cdots, l-1$; and b_1 is a least common multiple of the denominators of all elements of F, expressed in reduced form.[3]

Proof. Let y denote an l.c.m. of the denominators of all elements of F, expressed in reduced form. Then $yF \in \mathbf{R}^{n\times m}$, and the rank of yF is also l. By Theorem B.2.6, there exist unimodular matrices $U \in \mathbf{R}^{n\times n}$, $V \in \mathbf{R}^{m\times m}$ such that

$$
UyFV = \begin{bmatrix}
h_1 & \cdots & 0 & \\
\vdots & \vdots & \vdots & 0 \\
0 & \cdots & h_l & \vdots \\
& & 0 & 0
\end{bmatrix},
\tag{B.23}
$$

where h_i divides h_{i+1} for all i. Hence,

$$
UFV = \begin{bmatrix}
h_1/y & \cdots & 0 & \\
\vdots & \vdots & \vdots & 0 \\
0 & \cdots & h_l/y & \vdots \\
& & 0 & 0
\end{bmatrix}.
\tag{B.24}
$$

Let a_i/b_i be a reduced form for the fraction h_i/y, for $i = 1, \cdots, l$. Since $h_i \mid h_{i+1}$, let $h_{i+1} = h_i r_i$ where $r_i \in \mathbf{R}$. Let $[\cdot]$ denote the reduced form of a matrix. Then

$$
\frac{a_{i+1}}{b_{i+1}} = \left[\frac{h_{i+1}}{y}\right] = \left[\frac{h_i r_i}{y}\right] = \left[\frac{a_i r_i}{b_i}\right] = a_i\left[\frac{r_i}{b_i}\right],
\tag{B.25}
$$

where in the last step we used the fact that a_i, b_i are coprime. Now (B.25) implies that $a_i \mid a_{i+1}$, $b_{i+1} \mid b_i$ for all i.

[3] See Fact A.3.6 for the definition of a reduced form.

Finally, to show that $b_1 \sim y$, it is enough to show that h_1 and y are coprime, since a_1/b_1 is a reduced form of h_1/y. This is most easily done using prime factorizations. Let p_{ij}/q_{ij} be a reduced form of the element f_{ij}, for all i, j. Then y is an l.c.m. of all q_{ij}, and h_1 is a GCD of all $yf_{ij} = yp_{ij}/q_{ij}$ for all i, j. Suppose t is a prime factor of y, of multiplicity of α. (By this we mean that t is a prime, t^α divides y but $t^{\alpha+1}$ does not.) Then, from Fact A.3.10, t must be a prime divisor of multiplicity α, of *some* q_{ij}. The corresponding p_{ij} is not divisible by t, since p_{ij}, q_{ij} are coprime (see Corollary A.3.9). As a result, the corresponding term $yf_{ij} = yp_{ij}/q_{ij}$ is also not divisible by t (since t is a factor of multiplicity α of both yp_{ij} and q_{ij}). Thus, h_1, being a GCD of all yf_{ij}, is also not divisible by t (see Fact A.3.8). Since this is true of *every* prime factor of y, it follows from Corollary A.3.9 that y and h_1 are coprime. $\qquad \square$

APPENDIX C

Topological Preliminaries

In this appendix, a few basic concepts from topology are introduced. For greater detail, the reader is referred to [56] or [89].

C.1 TOPOLOGICAL SPACES

This section contains a brief introduction to topological spaces.

Definition C.1.1 Let S be a set. A collection T of subsets of S is a *topology* if[1]
(TOP1) Both S and $\emptyset$ (the empty set) belong to T.
(TOP2) A finite intersection of sets in T again belongs to T.
(TOP3) An arbitrary union of sets in T again belongs to T.
The ordered pair (S, T) is called a *topological space*, and subsets of S belonging to T are said to be *open*. A subset of S is *closed* if its complement in S is open.

Examples C.1.2 Let S be any set, and let T_1 consist of just the two sets S and $\emptyset$. Then (S, T_1) is a topological space. T_1 is referred to as the *trivial* topology on S. Let T_2 consist of all subsets of S. Then (S, T_2) is also a topological space. T_2 is referred to as the *discrete* topology on S.

Suppose S is a set, and T_1, T_2 are topologies on S. Then T_1 is *weaker* than T_2 (and T_2 is *stronger* than T_1) if T_1 is a subset of T_2, i.e., every set that is open in the topology T_1 is also open in the topology T_2. It is obvious that, for any set S, the trivial topology and the discrete topology are, respectively, the weakest and strongest topologies that can be defined on S.

To give interesting and useful examples of topological spaces, the notion of base is introduced. To motivate this notion, recall the familiar definition of an open subset of the real line: A subset U of $\mathbb{R}$ is open if and only if, corresponding to every $x \in U$, there is a number $\delta > 0$ such that the interval $(x - \delta, x + \delta)$ is also contained in U. The concept of a base is an abstraction of this idea, whereby intervals of the form $(x - \delta, x + \delta)$ are replaced by more general sets satisfying appropriate axioms.

Let S be a set, and let B be a collection of subsets of S that satisfies two axioms:
(B1) The sets in B cover S (i.e., the union of the sets in B is S).
(B2) Whenever B_1, B_2 are sets in B with a nonempty intersection and $x \in B_1 \cap B_2$, there exists a $B(x)$ in B such that $x \in B(x) \subseteq (B_1 \cap B_2)$.

[1]Throughout this section, upper case italic letters denote collections of sets, while bold face letters denote sets.

Using this collection B, another collection T of subsets of $\mathbf{S}$ is defined as follows: A subset $\mathbf{U}$ of $\mathbf{S}$ belongs to T if and only if, for every $x \in \mathbf{U}$, there is a set $\mathbf{B}(x)$ such that $x \in \mathbf{B}(x) \subseteq \mathbf{U}$.

Fact C.1.3   A subset $\mathbf{U}$ of $\mathbf{S}$ belongs to the collection T if and only if it is a union of sets in B. T is a topology on $\mathbf{S}$. Moreover, T is the weakest topology on $\mathbf{S}$ containing all sets in B.

Remarks C.1.4 B is referred to as a *base* for the topology T, and T is the *topology generated* by the base B.

Example C.1.5   To clarify Fact C.1.3, consider the set $\mathbb{R}^n$, consisting of n-tuples of real numbers. Let B denote the collection of "balls" $\mathbf{B}(x, \varepsilon)$ of the form

$$\mathbf{B}(x, \varepsilon) = \{y \in \mathbb{R}^n : \|x - y\| < \varepsilon\} \tag{C.1}$$

as x varies over $\mathbb{R}^n$ and ε varies over the positive numbers. Here $\| \cdot \|$ denotes the usual Euclidean (or any other) norm on $\mathbb{R}^n$. It is a straight-forward matter to verify that this collection of balls satisfies axioms (B1) and (B2). Hence, this collection forms a base for a topology on $\mathbb{R}^n$, in which a set $\mathbf{U} \subseteq \mathbb{R}^n$ is open if and only if, for every $x \in \mathbf{U}$, there is a ball $\mathbf{B}(x, \varepsilon) \subseteq \mathbf{U}$. This coincides with the "usual" definition of open sets on $\mathbb{R}^n$. The one extra bit of information that comes out of Fact C.1.3 is that a set is open if and only if it is a union of balls.

Proof of Fact C.1.3.   To prove the first sentence, suppose first that $\mathbf{U}$ is a union of sets in B; it is shown that $\mathbf{U}$ belongs to the collection T. Specifically, suppose $\mathbf{U} = \bigcup_{i \in I} \mathbf{B}_i$, where I is an index set and $\mathbf{B}_i$ is in the collection B for all $i \in I$. Let x be an arbitrary element of $\mathbf{U}$. Then $x \in \mathbf{B}_i$ for some $i \in I$, and for this i we have $x \in \mathbf{B}_i \subseteq \mathbf{U}$. Hence, by the definition of T, $\mathbf{U}$ is in T. Conversely, suppose $\mathbf{U}$ is in T. Then for every $x \in \mathbf{U}$ there is a $\mathbf{B}(x)$ in B such that $x \in \mathbf{B}(x) \subseteq \mathbf{U}$. It is now claimed that $\mathbf{U} = \bigcup_{x \in \mathbf{U}} \mathbf{B}(x)$, which would show that $\mathbf{U}$ is a union of sets from B. To prove the claim (and thereby complete the proof of the first sentence), note that $\mathbf{U} \supseteq \bigcup_{x \in \mathbf{U}} \mathbf{B}(x)$ since each $\mathbf{B}(x) \subseteq \mathbf{U}$; conversely, since every $x \in \mathbf{U}$ also belongs to the corresponding $\mathbf{B}(x)$, it follows that $\mathbf{U} \subseteq \bigcup_{x \in \mathbf{U}} \mathbf{B}(x)$. Hence, the two sets are equal.

The second sentence is proved by verifying that T satisfies the three axioms (TOP1)–(TOP3). First, $\varnothing$ belongs to T since $\varnothing$ vacuously satisfies the defining condition for a set to belong to T. Next, since $\mathbf{S}$ is the union of all sets in B, it is also in T. Hence, T satisfies (TOP1).

To establish (TOP2), it is enough to show that a nonempty intersection of two sets in T again belongs to T; it will then follow by induction that every *finite* intersection of sets in T again belongs to T. Accordingly, suppose $\mathbf{U}, \mathbf{V}$ are in T and let x be any element of $\mathbf{U} \cap \mathbf{V}$; we will show the existence of a $\mathbf{B}(x)$ in B such that $x \in \mathbf{B}(x) \subseteq (\mathbf{U} \cap \mathbf{V})$, which in turn will establish that the intersection

$\mathbf{U} \cap \mathbf{V}$ is open. Since $x \in \mathbf{U}$ and $x \in \mathbf{V}$, there exist $\mathbf{B}_1(x)$, $\mathbf{B}_2(x)$ in B such that $x \in \mathbf{B}_1(x) \subseteq \mathbf{U}$, $x \in \mathbf{B}_2(x) \subseteq \mathbf{V}$. Now, by axiom (B2), there is a $\mathbf{B}(x)$ in B such that $x \in \mathbf{B}(x) \subseteq (\mathbf{B}_1(x) \cap \mathbf{B}_2(x))$, which in turn is contained in $\mathbf{U} \cap \mathbf{V}$. Thus, T satisfies (TOP2).

To establish (TOP3), suppose $\{\mathbf{U}_\alpha\}$ is a family of sets in T. Then each $\mathbf{U}_\alpha$ is a union of sets belonging to B, whence their union is also a union of sets belonging to B. Thus, $\bigcup_\alpha \mathbf{U}_\alpha$ is also in T.

To prove the last sentence, let T_1 be another topology on $\mathbf{S}$ such that every set in B is contained in T_1. Since T_1 is a topology, arbitrary unions of sets in B again belong to T_1. Since every set in T can be expressed as a union of sets in B, it follows that every set in T is in T_1, i.e., T is weaker than T_1. $\qquad\qquad\square$

A topological space $(\mathbf{S}, T)$ is *first-countable* if T has the following property: For every $x \in \mathbf{S}$, there is a *countable* collection of open sets $\{\mathbf{B}_i(x), i \in \mathbf{Z}_+\}$, each containing x, such that every open set containing x also contains some $\mathbf{B}_i(x)$. In view of Fact C.1.3, the collection $\mathbf{B}_i(x)$, $x \in \mathbf{S}$ is a base for the topology T. Since the set $\mathbf{S}$ may be uncountable, the base for the topology may also be uncountable. However, in a first-countable topology, the collection of open sets *containing each particular point* has a countable base.

For example, consider the set $\mathbb{R}^n$ together with the topology of Example C.1.5. For a fixed $x \in \mathbb{R}^n$, the collection of balls $\mathbf{B}(x, 1/m)$, m an integer ≥ 1 is countable; moreover, every open set containing x also contains at least one of the balls $\mathbf{B}(x, 1/m)$. Hence, the topological space of Example C.1.5 is first-countable.

A very general class of first-countable topological spaces is that of metric spaces, which are defined next.

Definition C.1.6   A *metric space* $(\mathbf{S}, \rho)$ is a set $\mathbf{S}$, together with a function $\rho : \mathbf{S} \to \mathbb{R}$ satisfying the following axioms:

(M1) $\rho(y, x) = \rho(x, y) \, \forall x, y \in \mathbf{S}$.

(M2) $\rho(x, y) \geq 0 \, \forall x, y \in \mathbf{S}$; $\rho(x, y) = 0 \Longleftrightarrow x = y$.

(M3) $\rho(x, z) \leq \rho(x, y) + \rho(y, z) \, \forall x, y, z \in \mathbf{S}$.

If $(\mathbf{S}, \rho)$ is a metric space, then there is a natural topology that can be defined on $\mathbf{S}$. Let $\mathbf{B}(x, \varepsilon)$ denote the ball

$$\mathbf{B}(x, \varepsilon) = \{y \in \mathbf{S} : \rho(x, y) < \varepsilon\}. \qquad (C.2)$$

Then the collection of sets $\mathbf{B}(x, \varepsilon)$ as x varies over $\mathbf{S}$ and ε varies over all positive numbers, is a base for a topology T on $\mathbf{S}$. T is referred to as the topology on $\mathbf{S}$ *induced* by the metric ρ. In this topology, a set $\mathbf{U}$ in $\mathbf{S}$ is open if and only if, for every $x \in \mathbf{U}$, there is a ball $\mathbf{B}(x, \varepsilon) \subseteq \mathbf{U}$.

Every metric space is first-countable as a topological space: For a fixed x, consider the *countable* collection of sets $\mathbf{B}(x, 1/m)$, m an integer ≥ 1. Then every open set containing x also contains at least one of the sets $\mathbf{B}(x, 1/m)$.

The question of convergence of sequences in topological spaces is examined next. Suppose $(\mathbf{S}, T)$ is a topological space. A set $\mathbf{N}$ is said to be a *neighborhood* of $x \in \mathbf{S}$ if $x \in \mathbf{N}$, and $\mathbf{N}$ contains a set in T that contains x. In other words, a set $\mathbf{N}$ is a neighborhood of x if it contains an open set containing x. Note that a neighborhood itself need not be open. A sequence $\{x_i\}$ in $\mathbf{S}$ is said to *converge* to $x \in \mathbf{S}$ if every neighborhood of x contains all but a finite number of terms of the sequence $\{x_i\}$.

Example C.1.7   Consider the set $\mathbb{R}^n$ together with the topology defined in Example C.1.5. Then a sequence $\{x_i\}$ converges to x if and only if, for every $\varepsilon > 0$, there is a number N such that $x_i \in \mathbf{B}(x, \varepsilon) \, \forall i \geq N$. This is the familiar notion of convergence in $\mathbb{R}^n$.

Example C.1.8   Consider a set $\mathbf{S}$ together with the discrete topology of Example C.1.2. Let $\{x_i\}$ be a sequence in $\mathbf{S}$ converging to x. Since the singleton set $\{x\}$ is also a neighborhood of x, it follows that $x_i = x$ for all except a finite number of values of i.

Example C.1.9   Consider a set $\mathbf{S}$ together with the trivial topology of Example C.1.2. Then $\mathbf{S}$ is the only neighborhood of any $x \in \mathbf{S}$. As a result, every sequence $\{x_i\}$ converges to every $x \in \mathbf{S}$.

The preceding two examples show that the meaning of convergence is very much dependent on the particular topology defined on a set. Moreover, in general, a sequence does not necessarily have a unique limit. To address the latter problem, we define Hausdorff topologies. A topology T on a set $\mathbf{S}$ is *Hausdorff* if distinct points have disjoint neighborhoods. It is easy to verify that if T is a Hausdorff topology on a set $\mathbf{S}$, then the limit of a sequence in $\mathbf{S}$ is unique if it exists. Also, if $(\mathbf{S}, \rho)$ is a metric space, then the topology induced by the metric ρ is Hausdorff.

Let $\mathbf{S}$ be a set, and let T_1, T_2 be two topologies on $\mathbf{S}$, with T_1 weaker than T_2. Then, whenever a sequence $\{x_i\}$ in $\mathbf{S}$ converges to x in the topology T_2, it also converges to x in the topology T_1. However, it is possible that a sequence may converge in T_1 but not in T_2. This shows that convergence in T_1 is in general a weaker requirement than convergence in T_2 (and helps to explain the terminology).

Recall that a set in $\mathbf{S}$ is *closed* if its complement is open. One of the desirable features of first-countable topologies is that one can give an alternate, and very useful, characterization of closed sets. Suppose $\mathbf{U}$ is a set in $\mathbf{S}$. An element $x \in \mathbf{S}$ is said to be a *limit point* (or a cluster point, accumulation point) of $\mathbf{U}$ if every neighborhood of x contains an element of $\mathbf{U}$ other than x. It can be shown that a set is closed if and only if it contains all of its limit points. Now suppose $(\mathbf{S}, T)$ is a first-countable topological space. Then, for each $x \in \mathbf{S}$, there is a *countable* collection of open sets $\{\mathbf{B}_i(x), i \in \mathbf{Z}\}$ such that every neighborhood of x contains at least one of the $\mathbf{B}_i(x)$. As a result, x is a limit point of a set $\mathbf{U}$ if and only if there exists a *sequence* $\{x_i\}$ in $\mathbf{U}$ converging to x such that $x_i \neq x \, \forall i$. Based on this, one can show that $\mathbf{U}$ is closed if and only if every convergent sequence $\{x_i\}$ in $\mathbf{U}$ has a limit in $\mathbf{U}$ (see Problem C.1.5).

Now we come to the notion of continuity. Suppose $(\mathbf{S}_1, T_1)$, $(\mathbf{S}_2, T_2)$ are topological spaces, and f is a function mapping $\mathbf{S}_1$ into $\mathbf{S}_2$. Given any subset $\mathbf{U}$ of $\mathbf{S}_2$, its *preimage* under f is the subset of $\mathbf{S}_1$ denoted by $f^{-1}(\mathbf{U})$ and defined by

$$f^{-1}(\mathbf{U}) = \{x \in \mathbf{S}_1 : f(x) \in \mathbf{U}\} . \tag{C.3}$$

The function f is *continuous* at $x \in \mathbf{S}_1$ if, whenever $\mathbf{U}$ is an open subset of $\mathbf{S}_2$ containing $f(x)$, its preimage $f^{-1}(\mathbf{U})$ is an open subset of $\mathbf{S}_1$ (containing x). f is *continuous* if it is continuous at all $x \in \mathbf{S}_1$. Clearly, whether or not f is continuous is very much dependent on the topologies T_1 and T_2. For example, if T_1 is the discrete topology, then *every* function $f : \mathbf{S}_1 \to \mathbf{S}_2$ is continuous. Also, one can easily verify the following: If f is continuous at $x \in \mathbf{S}_1$ and $\mathbf{N}$ is a neighborhood of $f(x)$, then $f^{-1}(\mathbf{N})$ is a neighborhood of x (see Problem C.1.6). The converse is also true, but more difficult to prove: If $f^{-1}(\mathbf{N})$ is a neighborhood of x whenever $\mathbf{N}$ is a neighborhood of $f(x)$, then f is continuous at x. The proof can be found in [56].

Fact C.1.10   Suppose $(\mathbf{S}_1, T_1)$, $(\mathbf{S}_2, T_2)$ are topological spaces, and $f : \mathbf{S}_1 \to \mathbf{S}_2$ is continuous at $x \in \mathbf{S}_1$. Then, whenever $\{x_i\}$ is a sequence in $\mathbf{S}_1$ converging to x, the sequence $\{f(x_i)\}$ converges to $f(x)$.

Proof. Let $\mathbf{N}$ be any neighborhood of $f(x)$. Since f is continuous at x, $f^{-1}(\mathbf{N})$ is a neighborhood of x. Since $\{x_i\}$ converges to x, $f^{-1}(\mathbf{N})$ contains all but a finite number of terms of the sequence $\{x_i\}$. This is the same as saying that $f(x_i) \in \mathbf{N}$ for all but a finite number of values of i. Since this is true for *every* neighborhood of $f(x)$, we conclude that $\{f(x_i)\}$ converges to $f(x)$. $\square$

If the topologies T_1, T_2 are first-countable, then the converse of Fact C.1.10 is also true: If $\{f(x_i)\}$ converges to $f(x)$ for all sequences $\{x_i\}$ converging to x, then f is continuous at x.

PROBLEMS

C.1.1. Let $(\mathbf{S}, \rho)$ be a metric space, and let $\mathbf{B}(x, \varepsilon)$ be defined by (C.2). Show that the collection of sets $\mathbf{B}(x, \varepsilon)$ is a base for a topology on $\mathbf{S}$ by verifying that axiom (B1) is satisfied.

C.1.2. Let $\mathbf{S}$ be any set, and define $\rho : \mathbf{S} \times \mathbf{S} \to \mathbb{R}$ by $\rho(x, y) = 0$ if $x = y$, 1 if $x \neq y$.

(i) Verify that $(\mathbf{S}, \rho)$ is a metric space.

(ii) Show that the topology on $\mathbf{S}$ induced by ρ is the discrete topology.

C.1.3. Let $(\mathbf{S}, T)$ be a first countable topological space, and let $\mathbf{U}$ be a subset of $\mathbf{S}$. Show that $\mathbf{U}$ is closed if and only if the following is true: $x_i \in \mathbf{U}$, $\{x_i\}$ converges to x implies that $x \in \mathbf{U}$ (Hint: either $x = x_i$ for some i or else x is a limit point of $\mathbf{U}$).

C.1.4. Suppose $(\mathbf{S}, \rho)$ is a metric space, and define

$$d(x, y) = \frac{\rho(x, y)}{1 + \rho(x, y)} \ .$$

(i) Show that $(\mathbf{S}, d)$ is also a metric space.

(ii) Show that ρ and d induce the same topology on $\mathbf{S}$.

C.1.5. Show that, if $(\mathbf{S}, T)$ is a topological space and T is a Hausdorff topology, then a sequence in $\mathbf{S}$ can converge to at most one point.

C.1.6. Suppose $(\mathbf{S}_1, T_1)$, $(\mathbf{S}_2, T_2)$ are topological spaces, and $f : \mathbf{S}_1 \to \mathbf{S}_2$ is continuous at $x \in \mathbf{S}_1$. Show that, if $\mathbf{N}$ is a neighborhood of $f(x)$, then $f^{-1}(\mathbf{N})$ is a neighborhood of x.

C.2   TOPOLOGICAL RINGS AND NORMED ALGEBRAS

This section contains a brief introduction to the subject of topological rings. Roughly speaking, a topological ring is a ring, together with a topology, such that subtraction and multiplication are continuous operations with respect to the topology. Before talking about topological rings as such, it is necessary to define the product topology on the cartesian product of topological spaces.

Recall that the *cartesian product* $\mathbf{S}_1 \times \mathbf{S}_2$ of two sets $\mathbf{S}_1$ and $\mathbf{S}_2$ consists of all ordered pairs (x, y) where $x \in \mathbf{S}_1$ and $y \in \mathbf{S}_2$. Now suppose T_1, T_2 are topologies on $\mathbf{S}_1, \mathbf{S}_2$, respectively. The objective is to define a topology on the product set $\mathbf{S} = \mathbf{S}_1 \times \mathbf{S}_2$. This is done in the following way: Let B denote the collection of subsets of $\mathbf{S}$ of the form $\mathbf{U} \times \mathbf{V}$ where $\mathbf{U} \in T_1, \mathbf{V} \in T_2$. In other words, B is the collection of subsets of $\mathbf{S}$ formed by taking cartesian products of an open subset of $\mathbf{S}_1$ and an open subset of $\mathbf{S}_2$. It can be shown that B is a base for a topology on $\mathbf{S}$ (see Problem C.2.1), which is referred to as the *product topology* on $\mathbf{S}$.

Now suppose $\mathbf{S}_i, i = 1, 2, 3$ are topological spaces, with the topologies $T_i, i = 1, 2, 3$. Let $\mathbf{S} = \mathbf{S}_1 \times \mathbf{S}_2$, and let T denote the product topology on $\mathbf{S}$. Suppose f is a function mapping $\mathbf{S}$ into $\mathbf{S}_3$. By definition, f is continuous if $f^{-1}(\mathbf{W})$ is an open subset of $\mathbf{S}$ whenever $\mathbf{W}$ is an open subset of $\mathbf{S}_3$. Now suppose f is continuous, that $(x, y) \in \mathbf{S}$ and let $f(x, y) =: z \in \mathbf{S}_3$. Since f is continuous, it follows that, whenever $\mathbf{N}$ is a neighborhood of z, $f^{-1}(\mathbf{N})$ is a neighborhood of (x, y). Recalling the definition of the product topology on the set $\mathbf{S}$, we see that, whenever $\mathbf{N}$ is a neighborhood of z, there exist neighborhoods $\mathbf{N}_1$ of x and $\mathbf{N}_2$ of y such that $f(x_1, y_1) \in \mathbf{N} \, \forall \, x_1 \in \mathbf{N}_1, \forall \, y_1 \in \mathbf{N}_2$. In particular, $f(x, y_1) \in \mathbf{N} \, \forall \, y_1 \in \mathbf{N}_2$, and $f(x_1, y) \in \mathbf{N} \, \forall \, x_1 \in \mathbf{N}_1$. The conclusion is that if f is a continuous function, then for each $x \in \mathbf{S}_1$ the function $f(x, \cdot) : \mathbf{S}_2 \to \mathbf{S}_3$ is continuous; similarly, for each $y \in \mathbf{S}_2$, the function $f(\cdot, y) : \mathbf{S}_1 \to \mathbf{S}_3$ is continuous.

Now we come to topological rings.

Definition C.2.1   Suppose $\mathbf{R}$ is a ring and T is a topology on $\mathbf{R}$. The pair $(\mathbf{R}, T)$ is a *topological ring* if the functions $(x, y) \mapsto x - y$ and $(x, y) \mapsto xy$ are continuous functions from $\mathbf{R} \times \mathbf{R}$ into $\mathbf{R}$, when $\mathbf{R} \times \mathbf{R}$ is given the product topology.

Several facts are immediate consequences of the above definition. First, if the function $(x, y) \mapsto x - y$ is continuous, then *for each fixed x*, the function $y \mapsto x - y$ is a continuous function from $\mathbf{R}$ into $\mathbf{R}$. In particular, taking $x = 0$, it follows that the function $y \mapsto -y$ is continuous. Next, since compositions of continuous functions are again continuous, the function $(x, y) \mapsto x + y = x - (-y)$ is also continuous. Similarly, the function $(x, y) \mapsto y - x = -(x - y)$ is also continuous.

Given a subset $\mathbf{U} \subseteq \mathbf{R}$ and an element $x \in \mathbf{R}$, let $x + \mathbf{U}$ denote the set defined by

$$x + \mathbf{U} = \{x + y : y \in \mathbf{U}\}. \tag{C.1}$$

One can think of $x + \mathbf{U}$ as the set $\mathbf{U}$ "translated" by x. Now suppose $\mathbf{U}$ is an *open* subset of $\mathbf{R}$, and define the function $f_x : \mathbf{R} \to \mathbf{R}$ by $f_x(y) = y - x$. As seen earlier, f_x is continuous. Hence, the set $f_x^{-1}(\mathbf{U})$ is open. But clearly this set equals $x + \mathbf{U}$. Thus, we have shown that if $\mathbf{U}$ is open, so is $x + \mathbf{U}$ for all $x \in \mathbf{R}$. In other words, the topology on a topological ring is "translation-invariant," in the sense that translates of open sets are again open. A consequence of this is that, once we know all the open sets containing 0, we know all open sets.

We now speak briefly about *normed algebras*. Suppose $\mathbf{A}$ is a linear vector space over a field $\mathbf{F}$. This means that there is a concept of addition between members of $\mathbf{A}$, and scalar multiplication between an element of $\mathbf{F}$ and an element of $\mathbf{A}$. Now $\mathbf{A}$ is an algebra (over $\mathbf{F}$) if, in addition, one can define a product of two elements of $\mathbf{A}$ in such a way that $\mathbf{A}$ becomes a ring. Thus, an algebra $\mathbf{A}$ has associated with it a field $\mathbf{F}$, and *three* operations: addition between elements of $\mathbf{A}$, multiplication between two elements of $\mathbf{A}$, and multiplication between an element of $\mathbf{F}$ and an element of $\mathbf{A}$. $\mathbf{A}$ is a *ring* with respect to addition and multiplication, and $\mathbf{A}$ is a *linear vector space* with respect to addition and scalar multiplication. For example, if $\mathbf{F}$ is any field, the set of matrices $\mathbf{F}^{n \times n}$ is an algebra over $\mathbf{F}$.

Now suppose $\mathbf{A}$ is an algebra over $\mathbb{R}$ (the real numbers) or C (the complex numbers). Then $(\mathbf{A}, \|\cdot\|)$ is a *normed algebra* if one can define a norm $\|\cdot\|$ on $\mathbf{A}$ such that

(NA1) $\|a\| \geq 0 \, \forall \, a \in \mathbf{A}$, $\|a\| = 0 \iff a = 0$,

(NA2) $\|\alpha a\| = |\alpha| \cdot \|a\| \, \forall \, \alpha \in \mathbb{R}(\text{ or C}), \, \forall \, a \in \mathbf{A}$,

(NA3) $\|a + b\| \leq \|a\| + \|b\| \, \forall \, a, b \in \mathbf{A}$,

(NA4) $\|ab\| \leq \|a\| \cdot \|b\|$, $\forall \, a, b \in \mathbf{A}$,

Axioms (NA1)–(NA3) are just the usual requirements for $(\mathbf{A}, \|\cdot\|)$ to be a normed space, and (NA4) is the additional requirement for it to be a normed algebra.

Let $(\mathbf{A}, \|\cdot\|)$ be a normed algebra, and let B denote the collection of balls

$$\mathbf{B}(x, \varepsilon) = \{y \in \mathbf{A} : \|x - y\| < \varepsilon\}. \tag{C.2}$$

Then, B is a base for a topology T on $\mathbf{A}$. This is the same as the topology induced by the metric $\rho(x, y) = \|x - y\|$.

It is now shown that $(\mathbf{A}, T)$ is a topological ring. In order to do this, it is necessary to show that the functions $(x, y) \mapsto x - y$ and $(x, y) \mapsto xy$ are both continuous. Let us begin with subtraction. Suppose $\bar{x}, \bar{y} \in \mathbf{A}$, $\bar{z} = \bar{x} - \bar{y}$, and let $\mathbf{S} \subseteq \mathbf{A}$ be an open set containing $\bar{z}$. We must show that the set $\{(x, y) : x - y \in \mathbf{S}\} =: \mathbf{V}$ is an open subset of $\mathbf{A} \times \mathbf{A}$. This is done by showing that, whenever

$(x, y) \in \mathbf{V}$, there are balls $\mathbf{B}(x, \delta)$, $\mathbf{B}(y, \delta)$ such that $\mathbf{B} := \mathbf{B}(x, \delta) \times \mathbf{B}(y, \delta)$ is contained in $\mathbf{V}$. Since $\mathbf{S}$ is open and $x - y \in \mathbf{S}$, by definition there is an $\varepsilon > 0$ such that $\mathbf{B}(x - y, \varepsilon) \subseteq \mathbf{S}$. Now let $\delta = \varepsilon/2$, and, suppose $x_1 \in \mathbf{B}(\bar{x}, \delta)$, $y_1 \in \mathbf{B}(\bar{y}, \delta)$. Then

$$\|(x_1 - y_1) - (x - y)\| \le \|x_1 - x\| + \|y_1 - y\| < 2\delta < \varepsilon . \tag{C.3}$$

Hence, $(x_1, y_1) \in \mathbf{V}$. This shows that subtraction is continuous. Next, to show that multiplication is continuous, suppose $\bar{x}, \bar{y} \in \mathbf{A}$, let $\bar{z} = \bar{x}\bar{y}$, and let $\mathbf{S}$ be an open set in $\mathbf{A}$ containing $\bar{z}$. Define $\mathbf{V} = \{(x, y) : xy \in \mathbf{S}\}$. We show that $\mathbf{V}$ is open by showing that, whenever $(x, y) \in \mathbf{V}$, there exist balls $\mathbf{B}(x, \delta)$, $\mathbf{B}(y, \delta)$ such that $\mathbf{B}(x, \delta) \times \mathbf{B}(y, \delta)$ is contained in $\mathbf{V}$. Since $\mathbf{V}$ is open, by definition there is an $\varepsilon > 0$ such that $\mathbf{B}(xy, \varepsilon) \subseteq \mathbf{S}$. Suppose $x_1 \in \mathbf{B}(x, \delta)$, $y_1 \in \mathbf{B}(y, \delta)$ where δ is yet to be specified. Then

$$\begin{aligned} x_1 y_1 - xy &= [(x_1 - x) + x] \cdot [(y_1 - y) + y] - xy \\ &= (x_1 - x)(y_1 - y) + (x_1 - x)y + x(y_1 - y) . \end{aligned} \tag{C.4}$$

$$\begin{aligned} \|x_1 y_1 - xy\| &\le \|x_1 - x\| \cdot \|y_1 - y\| + \|x_1 - x\| \cdot \|y\| \\ &\quad + \|x\| \cdot \|y_1 - y\| \\ &\le \delta^2 + \delta(\|x\| + \|y\|) . \end{aligned} \tag{C.5}$$

Hence $(x_1, y_1) \in \mathbf{V}$ if δ is chosen such that

$$\delta^2 + \delta(\|x\| + \|y\|) \le \varepsilon . \tag{C.6}$$

Such a δ can always be found since the left side of (C.6) is continuous in δ and equals zero when δ is zero. This completes the proof that multiplication is continuous.

Fact C.2.2   Suppose $(\mathbf{A}, \|\cdot\|)$ is a normed algebra, and let $\mathbf{U}$ denote the set of units of $\mathbf{A}$ (i.e., the set of elements of $\mathbf{A}$ that have a multiplicative inverse in $\mathbf{A}$). Then the function $f : u \mapsto u^{-1}$ maps $\mathbf{U}$ into itself continuously.

Remark C.2.3   By the continuity of the "inversion" function f, we mean the following: Given any $u \in \mathbf{U}$ and any $\varepsilon > 0$, there exists a $\delta > 0$ such that

$$\|u^{-1} - v^{-1}\| < \varepsilon \text{ whenever } v \in \mathbf{U} \text{ and } \|u - v\| < \delta . \tag{C.7}$$

Proof. Let $u \in \mathbf{U}$ and let $\varepsilon > 0$ be specified. Suppose $v \in \mathbf{U}$ and $v \in \mathbf{B}(u, \delta)$, where δ is not as yet specified. Then

$$\|u^{-1} - v^{-1}\| = \|u^{-1}(v - u)v^{-1}\| =$$
$$\leq \|u^{-1}\| \cdot \|v - u\| \cdot \|v^{-1}\|$$
$$\leq \delta \|u^{-1}\| \cdot \|v^{-1}\|$$
$$\leq \delta \|u^{-1}\| \cdot (\|u^{-1}\| + \|u^{-1} - v^{-1}\|) . \tag{C.8}$$

Solving for $\|u^{-1} - v^{-1}\|$ from (C.8) gives

$$\|u^{-1} - v^{-1}\| \leq \frac{\delta \|u^{-1}\|^2}{1 - \delta \|u^{-1}\|} . \tag{C.9}$$

Thus, (C.7) is satisfied if δ is chosen such that the right side of (C.9) is less than ε. But such a δ can always be found, since the right side of (C.9) is continuous in δ and equals zero when δ is zero. This shows that the function f is continuous. $\square$

PROBLEMS

C.2.1. Suppose $(\mathbf{S}_1, T_1)$, $(\mathbf{S}_2, T_2)$ are topological spaces.

(i) Let T denote the collection of subsets of $\mathbf{S}_1 \times \mathbf{S}_2$ of the form $\mathbf{U} \times \mathbf{V}$ where $\mathbf{U} \in T_1, \mathbf{V} \in T_2$. Show that T is a base for a topology on the set $\mathbf{S}_1 \times \mathbf{S}_2$ (Hint: Show that T satisfies the axiom (B1) of Section B.1).

(ii) Suppose B_1, B_2 are respectively bases for the topologies T_1, T_2. Let B denote the collection of subsets of $\mathbf{S}_1 \times \mathbf{S}_2$ of the form $\mathbf{B}_1 \times \mathbf{B}_2$ where $\mathbf{B}_1 \in B_1, \mathbf{B}_2 \in B_2$. Show that B is also a base for a topology on $\mathbf{S}_1 \times \mathbf{S}_2$.

(iii) Show that B and T generate the same topology on $\mathbf{S}_1 \times \mathbf{S}_2$ (Hint: Show that every set in T is a union of sets in B).

C.2.2. Consider the ring $\mathbb{R}^{n \times n}$ of $n \times n$ matrices with real elements. Let $\| \cdot \|$ be any norm on the *vector space* $\mathbb{R}^n$, and define the norm of a *matrix* by

$$\|A\| = \sup_{x \neq 0} \frac{\|Ax\|}{\|x\|}$$

Show that $(\mathbb{R}^{n \times n}, \| \cdot \|)$ is a normed algebra.

C.2.3. Consider the ring $\mathbb{R}[s]$ of polynomials in the indeterminate s with real coefficients. Define the norm of $a(s) = \sum_i a_i s^i$ by

$$\|a(\cdot)\| = \sum_i |a_i| .$$

Show that $(\mathbb{R}[s], \| \cdot \|)$ is a normed algebra.

Bibliography

[1] V. Anantharam and C. A. Desoer, "On the stabilization nonlinear systems," *IEEE Trans. on Auto. Control,* **AC-29**, pp. 569–572, June 1984. DOI: 10.1109/TAC.1984.1103584 Cited on page(s)

[2] B. D. O. Anderson, "A note on the Youla-Bongiorno-Lu condition," Automatica, **12**, pp. 387–388, July 1976. DOI: 10.1016/0005-1098(76)90060-1 Cited on page(s) 46, 111

[3] B. D. O. Anderson and J. B. Moore, *Optimal Filtering*, Prentice-Hall, Englewood Cliffs, New Jersey, 1979. Cited on page(s)

[4] M. F. Atiyah and I. G. MacDonald, *Introduction to Commutative Algebra*, Addison-Wesley, Reading, MA., 1969. Cited on page(s)

[5] J. A. Ball and J. W. Helton, "A Beurling-Lax theorem for the Lie group $U(m, n)$ which contains most classical interpolation theory," *J. Operator Theory*, **9**, pp. 107–142, 1983. Cited on page(s)

[6] C. I. Byrnes, M. W. Spong and T. J. Tarn, "A several complex variables approach to feedback stabilization of neutral delay-differential systems," *Math. Sys. Thy.*, **17**, pp. 97–134, May 1984. DOI: 10.1007/BF01744436 Cited on page(s)

[7] F. M. Callier and C. A. Desoer, "Open-loop unstable convolution feedback systems with dynamical feedback," *Automatica*, **13**, pp. 507–518, Dec. 1976. DOI: 10.1016/0005-1098(76)90010-8 Cited on page(s)

[8] F. M. Callier and C. A. Desoer, "An algebra of transfer functions of distributed linear time-invariant systems," *IEEE Trans. Circ. and Sys.*, **CAS-25**, pp. 651–662, Sept. 1978. DOI: 10.1109/TCS.1978.1084544 Cited on page(s)

[9] F. M. Callier and C. A. Desoer, "Simplifications and clarifications on the paper 'An algebra of transfer functions of distributed linear time-invariant systems,'" *IEEE Trans. Circ. and Sys.*, **CAS-27**, pp. 320–323, Apr. 1980. DOI: 10.1109/TCS.1980.1084802 Cited on page(s)

[10] F. M. Callier and C. A. Desoer, "Stabilization, tracking and disturbance rejection in multivariable convolution systems," *Annales de la Societé Scientifique de Bruxelles*, **94**, pp. 7–51, 1980. Cited on page(s)

[11] F. M. Callier and C. A. Desoer, *Multivariable Feedback Systems*, Springer-Verlag, New York, 1982. Cited on page(s)

[12] B. C. Chang and J. B. Pearson, "Optimal Disturbance reduction in linear multivariable systems," *IEEE Trans. on Auto. Control*, **AC-29**, pp. 880–888, Oct. 1984. DOI: 10.1109/TAC.1984.1103409 Cited on page(s)

[13] M. J. Chen and C. A. Desoer, "Necessary and sufficient conditions for robust stability of linear distributed feedback systems," *Int. J. Control*, **35**, pp. 255–267, 1982. DOI: 10.1080/00207178208922617 Cited on page(s)

[14] M. J. Chen and C. A. Desoer, "Algebraic theory of robust stability of interconnected systems," *IEEE Trans. on Auto. Control*, **AC-29**, pp. 511–519, June 1984. DOI: 10.1109/TAC.1984.1103572 Cited on page(s)

[15] J. H. Chow and P. V. Kokotovic, "Eigenvalue placement in two-time-scale systems," *Proc. IFAC Symp. on Large Scale Sys.*, Udine, Italy, pp. 321–326, June 1976. Cited on page(s)

[16] C.-C. Chu and J. C. Doyle, "On inner-outer and spectral factorizations," *Proc. IEEE Conf. on Decision and Control*, Las Vegas, Dec. 1984. Cited on page(s)

[17] H. T. Colebrooke, *Algebra with Arithmetic and Mensuration*, John Murray, London, 1817, reprinted by Dr. Martin Sandig OHg, Wiesbaden, W. Germany, 1973. Cited on page(s) xvii

[18] P. R. Delsarte, Y. Genin and Y. Kamp, "The Nevanlinna-Pick problem for matrix-valued functions," *SIAM J. Appl. Math*, **36**, pp. 47–61, Feb. 1979. DOI: 10.1137/0136005 Cited on page(s)

[19] C. A. Desoer and W. S. Chan, "The feedback interconnection of linear time-invariant systems," *J. Franklin Inst.*, **300**, pp. 335–351, 1975. DOI: 10.1016/0016-0032(75)90161-1 Cited on page(s) 111

[20] C. A. Desoer and M. J. Chen, "Design of multivariable feedback systems with stable plant," *IEEE Trans. on Auto. Control*, **AC-26**, pp. 408–415, April 1981. DOI: 10.1109/TAC.1981.1102594 Cited on page(s)

[21] C. A. Desoer and C. L. Gustafson, "Design of multivariable feedback systems with simple unstable plant," *IEEE Trans. on Auto. Control*, **AC-29**, pp. 901–908, Oct. 1984. DOI: 10.1109/TAC.1984.1103400 Cited on page(s)

[22] C. A. Desoer and C. L. Gustafson, "Algebraic theory of linear multivariable feedback systems,' *IEEE Trans. on Auto. Control*, **AC-29**, pp. 909–917, Oct. 1984. DOI: 10.1109/TAC.1984.1103410 Cited on page(s) 111

[23] C. A. Desoer and C. A. Lin, "Two-step compensation of nonlinear systems," *Systems and Control Letters*, **3**, pp. 41–46, June 1983. DOI: 10.1016/0167-6911(83)90036-1 Cited on page(s)

[24] C. A. Desoer and R. W. Liu, "Global parametrization of feedback systems with nonlinear plants," *Systems and Control Letters*, **1**, pp. 249–251, Jan. 1982. DOI: 10.1016/S0167-6911(82)80006-6 Cited on page(s)

[25] C. A. Desoer, R. W. Liu, J. Murray and R. Saeks, "Feedback system design: The fractional representation approach to analysis and synthesis," *IEEE Trans. Auto. Control*, **AC-25**, pp. 399–412, June 1980. DOI: 10.1109/TAC.1980.1102374 Cited on page(s) 46, 111

[26] C. A. Desoer and M. Vidyasagar, *Feedback Systems: Input-Output Properties*, Academic Press, New York, 1975. Cited on page(s) 12, 72

[27] J. C. Doyle, "Robustness of multiloop linear feedback systems," *Proc. 17th Conf. Decision and Control*, Ft. Lauderdale, pp. 12–18, 1979. Cited on page(s)

[28] J. C. Doyle, K. Glover, P. P. Khargonekar and B. A. Francis, "State space solutions to standard H_2 and H_∞ control problems," *IEEE Transactions on Automatic Control*, 34, 831-847, 1989. Cited on page(s) xiii

[29] J. C. Doyle, B. A. Francis and A. Tannenbaum, *Feedback Control Theory*, MacMillan, New York, 1991 Cited on page(s) xiii

[30] J. Doyle, "Analysis of feedback systems with structured uncertainties," *IEE Proc.*, Part D, **129**, pp. 242–250, Nov. 1982. Cited on page(s)

[31] J. C. Doyle, "Synthesis of robust controllers and filters," *Proc. IEEE Conference on Decision and Control*, pp. 109–114, 1983. DOI: 10.1109/CDC.1983.269806 Cited on page(s)

[32] J. C. Doyle and G. Stein, "Multivariable feedback design: Concepts for a classical/modern synthesis," *IEEE Trans. Auto. Control*, *AC-26*, pp. 4–16, Feb. 1981. DOI: 10.1109/TAC.1981.1102555 Cited on page(s)

[33] J. C. Doyle, J. E. Wall and G. Stein, "Performance and robustness resuits for structured uncertainty," *Proc. IEEE Conf. on Decision and Control*, pp. 628–636, 1982. Cited on page(s)

[34] P. L. Duren, *The Theory of H^p-spaces*, Academic Press, New York, 1970. Cited on page(s)

[35] B. A. Francis, "The multivariable servomechanism problem from the input-output viewpoint," *IEEE Trans. Auto. Control*, **AC-22**, pp. 322–328, June 1977. DOI: 10.1109/TAC.1977.1101501 Cited on page(s)

[36] B. A. Francis, "On the Wiener-Hopf approach to optimal feedback design," *Systems and Control Letters*, **2**, pp. 197–201, Dec. 1982. DOI: 10.1016/0167-6911(82)90001-9 Cited on page(s)

[37] B. A. Francis, *A Course in H_∞ Control Theory*, Lecture Notes in Control and Information Sciences, Volume 88, Springer-Verlag, Heidelberg, 1987. Cited on page(s) xiii

[38] B. A. Francis, J. W. Helton and G. Zames, "H^∞-optimal feedback controllers for linear multivariable systems," *IEEE Trans. on Auto. Control*, **AC-29**, pp. 888–900, Oct. 1984. DOI: 10.1109/TAC.1984.1103387 Cited on page(s)

[39] B. A. Francis and M. Vidyasagar, "Algebraic and topological aspects of the regulator problem for lumped linear systems," *Automatica*, **19**, pp. 87–90, Jan. 1983. DOI: 10.1016/0005-1098(83)90078-X Cited on page(s)

[40] B. A. Francis and G. Zames, "On optimal sensitivity theory for SISO feedback systems," *IEEE Trans. Auto. Control*, **AC-29**, pp. 9–16, Jan. 1984. DOI: 10.1109/TAC.1984.1103357 Cited on page(s) xiii

[41] F. R. Gantmacher, *Theory of Matrices*, Chelsea, New York. Cited on page(s) 132

[42] I. M. Gel'fand, D. Raikov and G. Shilov, *Commutative normed rings*, Chelsea, New York, 1964. Cited on page(s)

[43] B. K. Ghosh and C. I. Byrnes, "Simultaneous stabilization and simultaneous pole-placement by non-switching dynamic compensation," *IEEE Trans. Auto. Control*, **AC-28**, pp. 735–741, June 1983. DOI: 10.1109/TAC.1983.1103299 Cited on page(s)

[44] K. Glover, "All optimal Hankel-norm approximations of linear multivariable systems and their L^∞-error bounds," *Int. J. Control*, **39**, pp. 1115–1193, June 1984. DOI: 10.1080/00207178408933239 Cited on page(s)

[45] M. Green and D. J. N. Limebeer, *Linear Robust Control*, Prentice-Hall, Englewood Cliffs, New Jersey, 1995. Cited on page(s) xiii

[46] V. Guillemin and A. Pollack, *Differential Topology*, Prentice-Hall, Englewood Cliffs, NJ., 1979. Cited on page(s)

[47] C. L. Gustafson and C. A. Desoer, "Controller design for linear multivariable feedback systems with stable plant," *Int. J. Control*, **37**, pp. 881–907, 1983. DOI: 10.1080/00207178308933018 Cited on page(s)

[48] E. Hille and R. S. Phillips, *Functional Analysis and Semigroups*, Amer. Math. Soc., Providence, RI, 1957. Cited on page(s)

[49] K. Hoffman, *Banach Spaces of Analytic Functions*, Prentice-Hall, Englewood Cliffs, NJ., 1962. Cited on page(s)

[50] N. T. Hung and B. D. O. Anderson, "Triangularization technique for the design of multivariable control systems," *IEEE Trans. Auto. Control*, **AC-24**, pp. 455–460, June 1979. DOI: 10.1109/TAC.1979.1102052 Cited on page(s) 30

[51] C. A. Jacobson, "Some aspects of the structure and stability of a class of linear distributed systems," Robotics and Automation Lab. Rept. No. 31, Renn. Poly. Inst., Dept. of ECSE, May 1984. Cited on page(s)

[52] N. Jacobson, *Lectures in Abstract Algebra*, Vol. I, Van-Nostrand, New York. 1953. Cited on page(s) 113

[53] T. Kailath, *Linear Systems*, Prentice-Hall, Englewood Cliffs, NJ., 1980. Cited on page(s) 70, 74

[54] E. W. Kamen, P. P. Khargonekar and A. Tannenoaum, "A local theory of linear systems with noncommensurate time delays," submitted for publication. Cited on page(s)

[55] E. W. Kamen, P. P. Khargonekar and A. Tannenbaum, "Stabilization of time-delay systems using finite-dimensional compensators," *IEEE Trans. on Auto. Control*, **AC-30**, pp. 75–78, Jan. 1985. DOI: 10.1109/TAC.1985.1103789 Cited on page(s)

[56] J. L. Kelley, *General Topology*, Van-Nostrand, New York, 1955. Cited on page(s) 143, 147

[57] P. P. Khargonekar and A. Tannenbaum, "Noneuclidean metrics and the robust stabilization of systems with parameter uncertainty," *IEEE Trans. Auto. Control*, **AC-30**, pp. 1005-1013, Oct. 1985. DOI: 10.1109/TAC.1985.1103805 Cited on page(s)

[58] H. Kimura, "Robust Stabilizability for a class of transfer functions," *IEEE Trans. Auto. Control*, **AC-29**, pp. 788–793, Sept. 1984. DOI: 10.1109/TAC.1984.1103663 Cited on page(s)

[59] P. Koosis, *The Theory of H^p spaces*, Cambridge University Press, Cambridge, 1980. Cited on page(s)

[60] V. Kučera, *Discrete Linear Control: The Polynomial Equation Approach*, Wiley, New York, 1979. Cited on page(s) xvi

[61] T. Y. Lam, *Serre's Conjecture*, Lecture notes in Mathematics, No. 635, Springer-Verlag, Berlin, 1978. Cited on page(s)

[62] V. Ya. Lin, "Holomorphic fiberings and multivalued functions of elements of a Banach algebra," (English translation) *Funct. Anal. Appl.*, **37**, pp. 122–128, 1973. DOI: 10.1007/BF01078884 Cited on page(s)

[63] H. Lindel and W. Lutkebohmert, "Projektive modulen fiber polynomialen erweiterungen von potenzreihenalgebren," *Archiv der Math*, **28**, pp. 51–54, 1977. DOI: 10.1007/BF01223888 Cited on page(s)

[64] R. W. Liu and C. H. Sung, "Linear feedback system design," *Circ. Sys. and Sig. Proc.*, **2**, pp. 35–44, 1983. DOI: 10.1007/BF01598142 Cited on page(s)

[65] C. C. MacDuffee, *Theory of Matrices*, Chelsea, New York, 1946. Cited on page(s) 125, 129

[66] H. Maeda and M. Vidyasagar, "Some results on simultaneous stabilization," *Systems and Control Letters*, **5**, pp. 205-208, Sept. 1984. DOI: 10.1016/S0167-6911(84)80104-8 Cited on page(s) 111

[67] H. Maeda and M. Vidyasagar, "Infinite gain margin problem in multivariable feedback systems," *Automatica*, **22**, pp. 131-133, Jan. 1986. DOI: 10.1016/0005-1098(86)90115-9 Cited on page(s)

[68] A. S. Morse, "System invariants under feedback and cascade control," *Proc. Int. Conf. on Math. Sys. Thy.*, Udine, Italy, 1976. Cited on page(s) 30

[69] B. Sz-Nagy and C. Foias, *Harmonic Analysis of Operators on Hilbert Space*, Elsevier, New York, 1970. Cited on page(s)

[70] C. N. Nett, "The fractional representation approach to robust linear feedback design: A self-contained exposition," Robotics and Automation Lab. Rept. No. 30, Renn. Poly. Inst., Dept. of ECSE, May 1984. Cited on page(s)

[71] C. N. Nett, C. A. Jacobson and M. J. Balas, "Fractional representation theory: Robustness with applications to finite-dimensional control of a class of linear distributed systems," *Proc. IEEE Conf. on Decision and Control*, pp. 268–280, 1983. DOI: 10.1109/CDC.1983.269841 Cited on page(s)

[72] C. N. Nett, C. A. Jacobson and M. J. Balas, "A connection between state-space and doubly coprime fractional representations," *IEEE Trans. Auto. Control*, **AC-29**, pp. 831–832, Sept. 1984. DOI: 10.1109/TAC.1984.1103674 Cited on page(s) 70

[73] G. Newton, L. Gould and J. F. Kaiser, *Analytic Design of Linear Feedback Controls*, Wiley, New York, 1957. Cited on page(s)

[74] L. Pernebo, "An algebraic theory for the design of controllers for linear multivariable systems," *IEEE Trans. Auto. Control*, **AC-26**, pp. 171–194, Feb. 1981. DOI: 10.1109/TAC.1981.1102554 Cited on page(s) 30

[75] V. P. Potapov, "The Multiplicative structure of J-contractive matrix functions," *Translations of the Amer. Math. Soc.*, **15**, pp. 131–243. 1960. Cited on page(s)

[76] D. Quillen, "Projective modules over polynomial rings," *Invent. Math.*, **36**, pp. 167–171, 1976. DOI: 10.1007/BF01390008 Cited on page(s)

[77] V. R. Raman and R. W. Liu, "A necessary and sufficient condition for feedback stabilization in a factorial ring," *IEEE Trans. Auto. Control*, **AC-29**, pp. 941–942, Oct. 1984. DOI: 10.1109/TAC.1984.1103395 Cited on page(s)

[78] H. H. Rosenbrock, *State Space and Multivariable Theory*, Nelson, London, 1970. Cited on page(s) 75

[79] J. J. Rotman, *An Introduction to Homological Algebra*, Academic Press, New York, 1979. Cited on page(s)

[80] W. Rudin, *Fourier Analysis on Groups*, John Wiley, New York, 1962. Cited on page(s)

[81] W. Rudin, *Real and Complex Analysis*, McGraw-Hill, New York, 1974. Cited on page(s) 13

[82] R. Saeks and J. Murray, "Feedback System Design: The tracking and disturbance rejection problems," *IEEE Trans. Auto. Control*, **AC-26**, pp. 203–218, Feb. 1981. DOI: 10.1109/TAC.1981.1102561 Cited on page(s)

[83] R. Saeks and J. Murray, "Fractional representation, algebraic geometry and the simultaneous stabilization problem," *IEEE Trans. Auto. Control*, **AC-27**, pp. 895–903, Aug. 1982. DOI: 10.1109/TAC.1982.1103005 Cited on page(s) 111

[84] R. Saeks, J. Murray, O. Chua, C. Karmokolias and A. Iyer, "Feedback system design: The single variate case - Part I," *Circ., Sys. and Sig. Proc.*, **1**, pp. 137–170, 1982. DOI: 10.1007/BF01600050 Cited on page(s) 111

[85] R. Saeks, J. Murray, O. Chua, C. Karmokolias and A. Iyer, "Feedback system design: The single variate case - Part II," *Circ., Sys. and Sig. Proc.*, **2**, pp. 3–34, 1983. DOI: 10.1007/BF01598141 Cited on page(s) 111

[86] D. Sarason, "Generalized interpolation in H^∞," *Trans. Amer. Math. Soc.*, **127**, pp. 179–203, May 1967. DOI: 10.2307/1994641 Cited on page(s)

[87] D. D. Siljak, "On reliability of control," *Proc. IEEE Conf. Decision and Control*, pp. 687–694, 1978. Cited on page(s) 111

[88] D. D. Siljak, "Dynamic reliability using multiple control systems," *Int. J. Control*, **31**, pp. 303–329, 1980. DOI: 10.1080/00207178008961043 Cited on page(s) 111

[89] G. Simmons, *Introduction to Topology and Modern Analysis*, McGraw-Hill, New York, 1966. Cited on page(s) 143

[90] D. E. Smith, *History of Mathematics*. Vol. I, Dover, New York, 1958. Cited on page(s) xvii

[91] A. A. Suslin, "Projective modules over a polynomial ring are free," *Soviet Math. Doklady*, **17**, pp. 1160–1164, 1976. Cited on page(s)

[92] M. Vidyasagar, "Input-output stability of a broad class of linear time-invariant multivariable feedback systems," *SIAM J. Control*, **10**, pp. 203–209, Feb. 1972. DOI: 10.1137/0310015 Cited on page(s) xv

[93] M. Vidyasagar, "Coprime factorizations and the stability of muitivariable distributed feedback systems," *SIAM J. Control*, **13**, pp. 1144–1155, Nov. 1975. DOI: 10.1137/0313071   Cited on page(s)

[94] M. Vidyasagar, *Nonlinear Systems Analysis*, Prentice-Hall, Englewood Cliffs, NJ., 1978. Cited on page(s)

[95] M. Vidyasagar, "On the use of right-coprime factorizations in distributed feedback systems containing unstable subsystems," *IEEE Trans. Circ. and Sys.*, **CAS-25**, pp. 916–921, Nov. 1978. DOI: 10.1109/TCS.1978.1084408 Cited on page(s)

[96] M. Vidyasagar, "The graph metric for unstable plants and robustness estimates for feedback stability," *IEEE Trans. Auto. Control*, **AC-29**, pp. 403–418, May 1984. DOI: 10.1109/TAC.1984.1103547 Cited on page(s) 70

[97] M. Vidyasagar and N. K. Bose, "Input-output stability of linear systems defined over measure spaces," *Proc. Midwest Symp. Circ. and Sys.*, Montreal, Canada, pp. 394–397, Aug. 1975. Cited on page(s)

[98] M. Vidyasagar and K. R. Davidson, "A parametrization of all stable stabilizing compensators for single-input-output systems," (under preparation). Cited on page(s) 30, 46

[99] M. Vidyasagar and H. Kimura, "Robust controllers for uncertain linear multivariable systems," *Automatica*, **22**, pp. 85-94, Jan. 1986. DOI: 10.1016/0005-1098(86)90107-X Cited on page(s)

[100] M. Vidyasagar and B. C. Lévy, "On the genericity of simultaneous stabilizability," *Symp. Math. Thy. of Netw. and Sys.*, Stockholm, 1985. Cited on page(s)

[101] M. Vidyasagar, H. Schneider and B. A. Francis, "Algebraic and topological aspects of feedback stabilization," *IEEE Trans. Auto. Control*, **AC-27**, pp. 880–894, Aug. 1982. DOI: 10.1109/TAC.1982.1103015 Cited on page(s) 111

[102] M. Vidyasagar and N. Viswanadham, "Algebraic design techniques for reliable stabilization," *IEEE Trans. Auto. Control*, **AC-27**, pp. 1085–1095, Oct. 1982. DOI: 10.1109/TAC.1982.1103086 Cited on page(s) 46, 70, 111

[103] M. Vidyasagar and N. Viswanadham, "Reliable stabilization using a multi-controller confiqutation," *Proc. IEEE Conf. on Decision and Control*, pp. 856–859, 1983. DOI: 10.1016/0005-1098(85)90008-1 Cited on page(s) 111

[104] B. L. van der Waerden, *Geomety and Algebra in Ancient Civilizations*, Springer Verlag, New York, 1984. Cited on page(s) xvii

[105] J. L. Walsh, *Interpolation and Approximation by Rational Functions in the Complex Domain*, AMS Colloqium Publ., Providence, RI., 1935. Cited on page(s)

[106] W. Wolovich, *Linear Multivariable Control*, Springer-Verlag, New York, 1974.   Cited on page(s) 70

[107] W. M. Wonham, *Linear Multivariable Control: A Geometric Approach*, 2nd ed., Springer-Verlag, New York, 1979. Cited on page(s)

[108] D. C. Youla, J. J. Bongiorno, Jr. and C. N. Lu, "Single-loop feedback stablization of linear multivariable plants," *Automatica*, **10**, pp. 159–173, 1974. DOI: 10.1016/0005-1098(74)90021-1 Cited on page(s) 30, 46, 111

[109] D. C. Youla, J. J. Bongiorno, Jr. and H. A. Jabr, "Modern Wiener-Hopf design of optimal controllers, Part I: The single-input case," *IEEE Trans. on Auto. Control*, **AC-21**, pp. 3–14, Feb. 1976. DOI: 10.1109/TAC.1976.1101139 Cited on page(s) 46

[110] D. C. Youla and G. Gnavi, "Notes on n-dimensional system theory," *IEEE Trans. Circ. and Sys.*, **CAS-26**, pp. 105–111, Feb. 1979. DOI: 10.1109/TCS.1979.1084614 Cited on page(s)

[111] D. C. Youla, H. A. Jabr and J. J. Bongiorno, Jr., "Modern Wiener-Hopf design of optimal controllers, Part II: The multivariable case," *IEEE Trans. Auto. Control*, **AC-21**, pp. 319–338, June 1976. DOI: 10.1109/TAC.1976.1101223 Cited on page(s) 111

[112] D. C. Youla and R. Pickel, "The Quillen-Suslin theorem and the structure of n-dimensional elementary polynomial matrices," *IEEE Trans. Circ. and Sys.*, **CAS-31**, pp. 513–518, June 1984. DOI: 10.1109/TCS.1984.1085545 Cited on page(s)

[113] G. Zames, "Feedback and optimal sensitivity: Model reference transformations, multiplicative seminorms and approximate inverses," *IEEE Trans. Auto. Control*, **AC-26**, pp. 301–320, April 1981. DOI: 10.1109/TAC.1981.1102603 Cited on page(s)

[114] G. Zames and A. El-Sakkary, "Unstable systems and feedback: The gap metric," *Proc. Allerton Conf.*, pp. 380–385, 1980. Cited on page(s)

[115] G. Zames and B. A. Francis, "A new approach to classical frequency methods: feedback and minimax sensitivity," *IEEE Trans. Auto. Control*, **AC-28**, pp. 585–601, May 1983. DOI: 10.1109/TAC.1983.1103275 Cited on page(s) xiii

[116] O. Zariski and P. Samuel, *Commutative Algebra*, Vol. I, Van-Nostrand, New York, 1958. Cited on page(s) 113, 122

[117] K. Zhou, J. C. Doyle and K. Glover, *Robust and Optimal Control*, Prentice-Hall, Englewood Cliffs, New Jersey, 1996. Cited on page(s) xiii

Author's Biography

MATHUKUMALLI VIDYASAGAR

Mathukumalli Vidyasagar was born in Guntur, India in 1947. He received the B.S., M.S., and Ph.D. degrees in electrical engineering from the University of Wisconsin in Madison, in 1965, 1967, and 1969, respectively. From the next twenty years he taught mostly in Canada, before returning to his native India in 1989. Over the next twenty years, he built up two research organizations from scratch, first the Centre for Artificial Intelligence and Robotics under the Ministry of Defence, Government of India, and later the Advanced Technology Center in Tata Consultancy Services (TCS), India's largest software company. After retiring from TCS in 2009, he joined the University of Texas at Dallas as the Cecil & Ida Green Chair in Systems Biology Science, and he became the Founding Head of the Bioengineering Department. His current research interests are stochastic processes and stochastic modeling, and their application to problems in computational biology. He has received a number of awards in recognition of his research, including the 2008 IEEE Control Systems Award.

Index

Printed in the United States
by Baker & Taylor Publisher Services